Lecture Notes in Applied and Computational Mechanics

Volume 30

Series Editors

Prof. Dr.-Ing. Friedrich Pfeiffer
Prof. Dr.-Ing. Peter Wriggers

Lecture Notes in Applied and Computational Mechanics

Edited by F. Pfeiffer and P. Wriggers

Further volumes of this series found on our homepage: springer.com

Vol. 29: Schanz, M.; Steinbach, O. (Eds.)
Boundary Element Analysis
571 p. 2006 [978-3-540-47465-4]

Vol. 28: Helmig, R.; Mielke, A.; Wohlmuth, B.I. (Eds.)
Multifield Problems in Solid and Fluid Mechanics
571 p. 2006 [978-3-540-34959-4]

Vol. 27: Wriggers P., Nackenhorst U. (Eds.)
Analysis and Simulation of Contact Problems
395 p. 2006 [978-3-540-31760-9]

Vol. 26: Nowacki, J.P.
Static and Dynamic Coupled Fields in Bodies
with Piezoeffects or Polarization Gradient
209 p. 2006 [978-3-540-31668-8]

Vol. 25: Chen C.-N.
Discrete Element Analysis Methods
of Generic Differential Quadratures
282 p. 2006 [978-3-540-28947-0]

Vol. 24: Schenk, C.A., Schuëller. G
Uncertainty Assessment of Large
Finite Element Systems
165 p. 2006 [978-3-540-25343-3]

Vol. 23: Frémond M., Maceri F. (Eds.)
Mechanical Modelling and Computational Issues
in Civil Engineering
400 p. 2005 [978-3-540-25567-3]

Vol. 22: Chang C.H.
Mechanics of Elastic Structures with Inclined Members:
Analysis of Vibration, Buckling and Bending of X-Braced
Frames and Conical Shells
190 p. 2004 [978-3-540-24384-7]

Vol. 21: Hinkelmann R.
Efficient Numerical Methods and Information-Processing
Techniques for Modeling Hydro- and Environmental
Systems
305 p. 2005 [978-3-540-24146-1]

Vol. 20: Zohdi T.I., Wriggers P.
Introduction to Computational Micromechanics
196 p. 2005 [978-3-540-22820-2]

Vol. 19: McCallen R., Browand F., Ross J. (Eds.)
The Aerodynamics of Heavy Vehicles:
Trucks, Buses, and Trains
567 p. 2004 [978-3-540-22088-6]

Vol. 18: Leine, R.I., Nijmeijer, H.
Dynamics and Bifurcations
of Non-Smooth Mechanical Systems
236 p. 2004 [978-3-540-21987-3]

Vol. 17: Hurtado, J.E.
Structural Reliability: Statistical Learning Perspectives
257 p. 2004 [978-3-540-21963-7]

Vol. 16: Kienzler R., Altenbach H., Ott I. (Eds.)
Theories of Plates and Shells:
Critical Review and New Applications
238 p. 2004 [978-3-540-20997-3]

Vol. 15: Dyszlewicz, J.
Micropolar Theory of Elasticity
356 p. 2004 [978-3-540-41835-1]

Vol. 14: Frémond M., Maceri F. (Eds.)
Novel Approaches in Civil Engineering
400 p. 2003 [978-3-540-41836-8]

Vol. 13: Kolymbas D. (Eds.)
Advanced Mathematical and Computational
Geomechanics
315 p. 2003 [978-3-540-40547-4]

Vol. 12: Wendland W., Efendiev M. (Eds.)
Analysis and Simulation of Multifield Problems
381 p. 2003 [978-3-540-00696-1]

Vol. 11: Hutter K., Kirchner N. (Eds.)
Dynamic Response of Granular and Porous Materials
under Large and Catastrophic Deformations
426 p. 2003 [978-3-540-00849-1]

Vol. 10: Hutter K., Baaser H. (Eds.)
Deformation and Failure in Metallic Materials
409 p. 2003 [978-3-540-00848-4]

Vol. 9: Skrzypek J., Ganczarski A.W. (Eds.)
Anisotropic Behaviour of Damaged Materials
366 p. 2003 [978-3-540-00437-0]

Vol. 8: Kowalski, S.J.
Thermomechanics of Drying Processes
365 p. 2003 [978-3-540-00412-7]

Vol. 7: Shlyannikov, V.N.
Elastic-Plastic Mixed-Mode Fracture Criteria
and Parameters
246 p. 2002 [978-3-540-44316-2]

Vol. 6: Popp K., Schiehlen W. (Eds.)
System Dynamics and Long-Term Behaviour
of Railway Vehicles, Track and Subgrade
488 p. 2002 [978-3-540-43892-2]

Vol. 5: Duddeck, F.M.E.
Fourier BEM: Generalization
of Boundary Element Method by Fourier Transform
181 p. 2002 [978-3-540-43138-1]

Micromechanics
of Contact
and Interphase Layers

Stanisław Stupkiewicz

 Springer

STANISŁAW STUPKIEWICZ
Institute of Fundamental Technological Research
Polish Academy of Sciences
Świętokrzyska 21
00-049 Warszawa
Poland
sstupkie@ippt.gov.pl

With 76 Figures

Library of Congress Control Number: 2006937129

ISSN 1613-7736

ISBN-10 3-540-49716-1 **Springer Berlin Heidelberg New York**
ISBN-13 978-3-540-49716-5 **Springer Berlin Heidelberg New York**

Springer is a part of Springer Science+Business Media
springer.com

© Springer-Verlag Berlin Heidelberg 2007
Printed in Germany

Typesetting: Data conversion by the author.
Final processing by PTP-Berlin Protago-TEX-Production GmbH, Germany (www.ptp-berlin.com)
Cover-Design: WMXDesign GmbH, Heidelberg
Printed on acid-free paper   89/3141/Yu - 5 4 3 2 1 0

Preface

This book addresses selected aspects of the broad area of micromechanics of interfaces and interface layers. Its restricted scope results from its origin: this book is an updated version of my habilitation thesis [132] which first appeared in 2005 in the series *IFTR Reports* published by the Institute of Fundamental Technological Research (IPPT) in Warsaw, Poland. As such, it covers only the topics related to my recent research interests.

I wish to mention here two of my teachers who largely influenced these interests and who also contributed to the results reported in this book. Professor Zenon Mróz was my first teacher at IPPT and the supervisor of my PhD thesis in 1996. He guided me into the field of modelling of contact phenomena, and he continuously advises me (and not only me) to look for simple models that capture the essence of the problem. Professor Henryk Petryk, also from IPPT, introduced me to the fascinating subject of micromechanics of martensitic phase transformations in shape memory alloys. I benefit from, and I also enjoy, the stimulating cooperation with him.

I wish to thank Petr Šittner and Vaclav Novák from the Institute of Physics in Prague for providing me with their results of compression tests of CuAlNi single crystals as well as for many inspiring discussions on the physical aspects of martensitic transformations. I thank my colleagues from IPPT – Stanisław Kucharski, Grzegorz Starzyński and Anna Bartoszewicz – for the measurements of contact compliance. I am indebted to Professor Bogdan Raniecki for his careful review of the manuscript of my habilitation thesis. His comments, suggested corrections, and numerous discussions helped me a lot in preparing the final version. Last but not least, I thank my wife Kasia for her continuous support.

Warsaw   *Stanisław Stupkiewicz*
September 2006

Contents

1

Introduction

Micromechanical analysis of heterogeneous materials allows prediction of their *macroscopic behaviour* from the known properties, microstructure, and interaction mechanisms of the constituents at the micro-scale. At the same time, it allows determination of the *microscopic response* at each point of a macroscopic body subjected to external loading. The micromechanical analysis involves thus, at least, two different scales of observation. The actual characteristic dimensions at these scales are, in fact, arbitrary, as long as they differ sufficiently, so that the two scales can be separated.

In the mechanics of continuous media, the mathematical notion of a surface, i.e. a two-dimensional manifold embedded in a three-dimensional space, is an idealization, i.e. a model, of a more complex reality. Depending on the scale of observation, the actual interface can be considered as a *surface* of zero thickness or as a *layer* of some finite thickness, possibly with a microstructure. The first point of view can then be regarded as macroscopic while the second one is microscopic.

At the macro-scale, the interface is thus seen as a surface, however, some properties of this surface or some properties related to this surface may depend on the microstructure of the interface layer and on the phenomena that occur at the micro-scale. The micromechanics of interfaces deals thus with the transition between these two scales, and its purpose is to determine the relations between the macroscopic and microscopic quantities.

The corresponding micro-macro transition applied to contact interactions of rough bodies is the main concern of the first part of this book. The microstructure of a contact layer is formed by surface asperities, but, in a sense, also by micro-inhomogeneity of strains and stresses induced by asperity interaction. The macroscopic contact response may involve phenomena such as contact compliance, friction, wear, lubrication, heat transfer, etc., described in terms of the corresponding macroscopic variables. The purpose of the micromechanical analysis is then to predict the macroscopic contact behaviour and to identify the influential macroscopic quantities.

Micromechanics of interfaces may also be understood differently, namely as the analysis, carried out at different scales, of materials that *contain* interfaces, and in which these interfaces significantly influence the behaviour of the material. This context is much closer to the classical homogenization, as the interfaces are common components of composite materials. However, it is not always the case that the local properties of the interfaces, or the interfacial phenomena, determine the macroscopic behaviour of a heterogeneous material. For example, as long as the constituents of a composite material are perfectly bonded, the interfaces separating the constituents have no effect on the effective properties of the composite. The situation changes when the interfaces have considerable thickness, or when debonding along the interfaces occurs, so that the deformations within the interfaces significantly contribute to the macroscopic deformation.

Pronounced effects can also be expected when the interfaces *propagate* through the material, as, for instance, in materials undergoing phase transformations. Propagation of phase transformation fronts is then associated with *evolution of microstructure*, so that the local phenomena at phase transformation fronts may, in fact, govern the macroscopic behaviour. Related phenomena are studied in the second part of this book, which is devoted to micromechanical analysis of evolving martensitic microstructures. Specifically, the micro-macro transition is carried out for laminated microstructures in shape memory alloys undergoing the stress-induced martensitic transformation.

As mentioned above, two application areas (contact of rough bodies and evolving martensitic microstructures) are addressed in this book. While the two areas are rather different, there are several reasons to discuss them together. First of all, in both cases, *interfaces* and interfacial phenomena are dealt with. Furthermore, in both cases, the analysis of interfaces involves the corresponding *layers*: either the layer is a microscopic counterpart of a macroscopic surface, as in the case of contact layers; or the interfaces appear as the entities separating the layers of parent and product phases, as in the case of martensitic microstructures. The specific problems analyzed in this book are also *complementary* in the sense that they address several important aspects of micromechanics of interfaces and layers: homogeneous and inhomogeneous layers; layers of finite and infinitesimal thickness; given and unknown interface orientations; microstructure evolution associated with propagation of interfaces. Finally, the *compatibility conditions*, see Sect. 2.4, appear in different forms in each specific problem analyzed in this work as an *essential* element of the respective description.

1.1 Objectives and Scope

The main objective of this monograph is to present micromechanical modelling tools that are suitable for the analysis of the class of problems discussed above. The importance and scientific relevance of this objective seems quite obvious, since micromechanics, due to its *predictive* capabilities, is a very attractive modelling approach which proved highly successful in many areas of mechanics. This book addresses several aspects and methodologies related to the micromechanics of contact and interphase layers, ranging from the rigorous method of asymptotic expansions to the practical application of the finite element method.

The importance of the topics addressed in this book stems also from the scientific and technological motivations of the specific applications. The contact phenomena of friction, wear, lubrication, etc., are very complex and still not fully understood. This is because of the vast diversity of contact pairs (each surface is characterized by possibly different material, roughness topography, and oxide, contaminant and lubricant layers) and contact conditions (contact pressure, sliding distance and velocity, temperature), e.g. Bowden and Tabor [16], Rabinowicz [109], Persson [100]. At the same time, friction controls or affects many processes, both in a desired and undesired way (c.f. car brakes and friction losses in engines, to mention just the most obvious examples), while the economic losses due to friction and wear have been estimated at 5% of the gross national product, cf. Feeny et al. [29]. Importantly, the complexity and importance of the related phenomena is in contrast with the simplicity of the models used to quantify them, such as the classical Coulomb law of friction and Archard's wear model.

Consider, for instance, metal forming processes, which naturally involve tool-workpiece contact interactions. In these processes, friction controls the material flow, again with positive and negative effects. On the other hand, contact phenomena affect surface finish of the product, wear determines the service life of tools, while the most efficient lubricants (e.g. chlorinated paraffin oils) have negative impact on the environment. The related problems of roughness evolution, lubricant film breakdown, pick-up and galling, cf. Olsson [89], constitute one of the motivations of the micromechanical analysis of contact layers, which is carried out below.

The analysis of contact layers presented in Chaps. 3–6 is focused on asperity interaction and on the related inhomogeneities of deformation within thin subsurface layers. The aim is to study the interaction of these inhomogeneities with the homogeneous macroscopic deformation and, specifically, to predict the effect of the macroscopic deformation on the contact response. While these topics are recognized as important in metal forming processes in which the macroscopic deformations are plastic, the related effects have not yet attracted sufficient attention in the case of the elasto-plastic asperity deformation regime.

The second part, Chaps. 7–9, is concerned with micromechanical modelling of shape memory alloys (SMA) undergoing stress-induced martensitic transformations. Shape memory alloys, with their unique behaviour, belong to the group of so-called *smart materials*. The interesting properties of shape memory alloys (shape memory effect, pseudoelasticity, and others) are related to development and evolution of microstructures which accompany the martensitic phase transformation, cf. Bhattacharya [13], Otsuka and Wayman [94]. Consequently, shape memory alloys are very attractive materials for advanced applications, such as micro-devices (e.g. pumps, engines, valves), actuators, joints, etc. Medical applications (blood vessel stents, orthodontic wires, and others) constitute another important application area of shape memory alloys. The need for accurate models of their thermomechanical behaviour is thus obvious.

The macroscopic behaviour of a polycrystalline SMA specimen is governed by transformations of atomic structure at the crystal lattice scale. A complete micromechanical model requires thus a multi-scale analysis accounting for the following microstructural elements and corresponding scales (starting from the lowest level): twins, martensitic plates, complex martensitic microstructures at the single crystal level, and crystal aggregates in a polycrystal. While attempts are made to develop such complete models, e.g. Patoor et al. [75, 96, 97], Thamburaja and Anand [149], there is still need for refined micromechanical modelling of the phenomena occurring at each of the scales involved. Clearly, the phenomenological modelling, preferably based on micromechanical considerations, e.g. Peultier et al. [105], Raniecki and Lexcellent [111, 112], is still the main approach to describe the complex macroscopic behaviour of SMA polycrystals.

This work is concerned with the micromechanical analysis of martensitic microstructures in SMA *single crystals*. A class of nested laminated microstructures is considered, and the micro-macro transition is performed for an *evolving microstructure*. The crystallography of transformation and distinct elastic anisotropy of the phases are fully accounted for in the corresponding micromechanical model. The related computational scheme is developed and, subsequently, applied to simulate the macroscopic response of SMA single crystals.

The content and the organization of this book are briefly outlined below. In Chap. 2, selected basic concepts, definitions and relationships are introduced, including the interior–exterior decomposition, compatibility conditions, and elements of homogenization. The two parts that follow, Chaps. 3–6 and Chaps. 7–9, are self-contained and thus can be read independently.

In Chap. 3, the mixed form of constitutive equations, applicable, for instance, for thin, homogeneous layers is introduced, and this formalism is next used to develop a phenomenological model of real contact area evolution in metal forming. In Chap. 4, the method of asymptotic expansions is applied to derive the equations of boundary layers which are induced by micro-inhomogeneous boundary conditions, e.g. by contact of rough bodies. In Chap. 5, micromechanical analysis of boundary layers is carried out by introducing a special averaging operation, and properties of the corresponding averages are derived. Finally, in Chap. 6, the finite element analysis of contact boundary layers is carried out. Asperity ploughing and flattening processes are analyzed, and the effect of macroscopic in-plane strain on the contact response is studied.

A brief introduction to martensitic microstructures in shape memory alloys is provided in Chap. 7 along with a short discussion of the microstructure and interfacial energy of the transition layer at the austenite–twinned martensite interface. In Chap. 8, a micromechanical model of evolution of laminated microstructures in single crystals undergoing stress-induced transformation is developed by combining micro-macro transition relations with a rate-independent phase transformation criterion. Macroscopic constitutive rate-equations are derived, and several applications are provided for Cu-based alloys undergoing the cubic-to-monoclinic and cubic-to-orthorhombic transformation. In Chap. 9, an approach is developed for the prediction of the microstructural parameters of stress-induced martensitic plates. The problem is formulated as a constrained minimization problem, and the approach is applied to study the effect of the stacking fault energy, load axis orientation, and temperature on microstructural parameters of internally faulted martensitic plates in a CuZnAl shape memory alloy.

1.2 Computational Tools

Practical applications of developed micromechanical models, including numerical computations, constitute an important part of this book. Detailed exposition of the related computational aspects is not provided. However, since these aspects have a significant influence on the overall success of implementation and application of the developed models, some remarks are provided below.

Symbolic algebra systems, such as *Mathematica* [163], provide tools for symbolic derivation of formulae, including symbolic differentiation. As long as complexity of expressions is moderate, *Mathematica* proves to be an efficient and convenient environment for combined symbolic and numerical computations. However, with increasing complexity of problems tackled, symbolic formulae grow in an uncontrolled way. This makes the symbolic systems unusable for problems such as derivation of the finite element formulae (element residual vector, tangent matrix, etc.) in nonlinear mechanics. The same problem was encountered in the course of numerical implementation of the micromechanical model of evolving laminated martensitic microstructures, cf. Chaps. 8 and 9. Accordingly, *AceGen* [66], a symbolic code generation system, has been used in order to overcome the problem of severe complexity of expressions and to efficiently derive the numerical codes.

The symbolic code generator *AceGen* is a *Mathematica* package which extends the algebraic and symbolic capabilities of *Mathematica* with the automatic differentiation technique, simultaneous optimization of expressions, and theorem proving by stochastic evaluation of expressions, cf. Korelc [68]. As a result, the usual problem of the uncontrolled growth of expressions can be avoided. At the same time, efficient and robust implementation is possible due to the high-level symbolic description employing the programming language of *Mathematica*. Accordingly, *AceGen* proves to be a very efficient tool for generation of numerical codes for problems of high complexity, such as nonlinear finite element codes including complex material models, advanced contact formulations, and sensitivity analysis, cf. Korelc [68], Krstulović-Opara et al. [73], Lengiewicz et al. [78], Stupkiewicz et al. [133].

Mathematica supplemented with *AceGen* was thus the main environment for the numerical simulations reported in Chaps. 8 and 9. Also the finite element codes used in Chap. 6 were derived using *AceGen*. The finite element computations of Chap. 6 were performed within the *Computational Templates* [67] environment which is closely integrated with *AceGen*.

2

Fundamentals of Micromechanics

Abstract: In this chapter, selected basic concepts, definitions, and relationships are introduced as a basis for further developments. The majority of the material as well as the exposition are quite standard and well recognized. However, the explicit micro-macro transition relations for simple laminates, provided in Sect. 2.6 and in Appendix A.4, are not easily found in the literature.

2.1 Notation

With the exception of Chap. 3 and Appendix A, where the matrix notation is used, the tensor notation is used throughout this work. When it is needed for clarity, the index notation is also provided and, by default, the summation rule over repeated indices is applied. The bold-face symbols are used for vectors and tensors, and basic tensor operations are listed in Table 2.1.

Table 2.1. Basic tensor operations: notation.

AB	$A_{ij}B_j$ or $A_{ij}B_{jk}$ or $A_{ijkl}B_{kl}$ or $A_{ijkl}B_{klmn}$
$\mathbf{A} \cdot \mathbf{B}$	$A_i B_i$ or $A_{ij}B_{ij}$
$\mathbf{A} \otimes \mathbf{B}$	$A_i B_j$ or $A_{ij}B_{kl}$
$\mathbf{A}^{\mathrm{T}}$	$(A^{\mathrm{T}})_{ij} = A_{ji}$ or $(A^{\mathrm{T}})_{ijkl} = A_{klij}$
$\mathbf{I}$	δ_{ij} (Kronecker delta)

The infinitesimal strain format is used throughout this work. As an exception, the kinematically exact form of the crystallographic theory of martensite, employing the finite deformation framework, is provided in Sect. 7.2.

2.2 Interior–Exterior Decomposition

A second-rank symmetric tensor $\mathbf{T}$ can be uniquely decomposed into its *interior* part $\mathbf{T_P}$ and *exterior* part $\mathbf{T_A}$ relative to a locally smooth surface represented by its normal $\mathbf{n}$, so that

$$\boxed{\mathbf{T} = \mathbf{T_P} + \mathbf{T_A}, \qquad \mathbf{T_P}\mathbf{n} = \mathbf{0}.} \tag{2.1}$$

The operators of this decomposition are the fourth-rank tensors $\mathbf{\Pi_P}$ and $\mathbf{\Pi_A}$,

$$\mathbf{T_P} = \mathbf{\Pi_P T}, \qquad \mathbf{T_A} = \mathbf{\Pi_A T}, \tag{2.2}$$

being functions of the normal $\mathbf{n}$. Their components in a Cartesian coordinate system are given by (Hill [47])

$$\begin{aligned}
(\Pi_\mathrm{P})_{ijkl} &= \tfrac{1}{2}(\delta_{ik} - n_i n_k)(\delta_{jl} - n_j n_l) + \tfrac{1}{2}(\delta_{jk} - n_j n_k)(\delta_{il} - n_i n_l), \\
(\Pi_\mathrm{A})_{ijkl} &= \tfrac{1}{2}(\delta_{ik} n_j n_l + \delta_{jk} n_i n_l + \delta_{il} n_j n_k + \delta_{jl} n_i n_k) - n_i n_j n_k n_l,
\end{aligned} \tag{2.3}$$

where δ_{ij} is the Kronecker delta.

In an intrinsic Cartesian coordinate system, such that components of the normal vector $\mathbf{n}$ are $(n_1, n_2, n_3) = (0, 0, 1)$, the components of $\mathbf{T_P}$ and $\mathbf{T_A}$ are

$$(T_\mathrm{P})_{ij} = \begin{pmatrix} T_{11} & T_{12} & 0 \\ T_{12} & T_{22} & 0 \\ 0 & 0 & 0 \end{pmatrix}, \qquad (T_\mathrm{A})_{ij} = \begin{pmatrix} 0 & 0 & T_{13} \\ 0 & 0 & T_{23} \\ T_{13} & T_{23} & T_{33} \end{pmatrix}, \tag{2.4}$$

where T_{ij} are the components of $\mathbf{T}$ in this intrinsic coordinate system. It is seen that operator $\mathbf{\Pi_A}$ preserves the components that contain an index 3 (out-of-plane components) and removes the others (in-plane components). Operator $\mathbf{\Pi_P}$ acts oppositely.

It can easily be verified that the following property holds for arbitrary symmetric tensors $\mathbf{S}$ and $\mathbf{T}$,

$$\mathbf{S_A} \cdot \mathbf{T_P} = 0. \tag{2.5}$$

The importance of the interior–exterior decomposition is clearly seen in Sect. 2.4 below, where the compatibility conditions at a bonded interface are compactly expressed in terms of interior and exterior parts of strain and stress tensors. We also note that, in the case of contact interface with $\mathbf{n}$ being the outward normal, the contact traction $\mathbf{t} = \boldsymbol{\sigma}\mathbf{n}$ is solely related to the exterior part of the stress tensor $\boldsymbol{\sigma}_\mathrm{A}$, namely

$$\mathbf{t} = \boldsymbol{\sigma}\mathbf{n} = \boldsymbol{\sigma}_\mathrm{A}\mathbf{n}. \tag{2.6}$$

This is also seen in terms of components in the intrinsic coordinate system,

$$t_i = \begin{pmatrix} t_{\mathrm{T}1} \\ t_{\mathrm{T}2} \\ t_\mathrm{N} \end{pmatrix}, \qquad \sigma_{ij} = \begin{pmatrix} \sigma_{11} & \sigma_{12} & t_{\mathrm{T}1} \\ \sigma_{12} & \sigma_{22} & t_{\mathrm{T}2} \\ t_{\mathrm{T}1} & t_{\mathrm{T}2} & t_\mathrm{N} \end{pmatrix}, \tag{2.7}$$

where t_N is the normal contact traction, while $t_{\mathrm{T}1}$ and $t_{\mathrm{T}2}$ are the components of the tangential contact traction (friction stress).

2.3 Elements of Homogenization

In this section, some basic concepts of the homogenization theory of hetero-geneous inelastic materials are briefly introduced in order to provide the re-ference for the developments in Chaps. 4, 5, and 8. The presented results are rather standard, and the respective derivations and proofs, omitted here, can be found in numerous works on homogenization, e.g. Aboudi [3] Hill [45, 46], Suquet [146], Willis [159]. The exposition below is mostly based on that of Suquet [146].

Let us consider a body made of a heterogeneous material. When the size of the heterogeneities is small compared to the size of the body, it is natural to introduce two different scales: the macro-scale and the micro-scale with, respectively, $\mathbf{x}$ and $\mathbf{y}$ being the corresponding spatial variables, and to esta-blish the macroscopic (effective) properties of the material at the macro-scale. This procedure, called *homogenization*, is only possible when the material is statistically homogeneous, i.e. when a *representative volume element* can be chosen. The choice of the r.v.e. is an important part of micromechanical mo-delling, and is strongly related to the homogenization approach used.

At a macroscopic point $\mathbf{x}$, two families of variables are considered: the macroscopic variables that describe the state of the equivalent homogeneous body at the macro-scale, and the microscopic ones which are related to the local states within the r.v.e. Below, the dependence of all these variables on $\mathbf{x}$ is not indicated explicitly.

Under the assumptions of displacement continuity and mechanical equili-brium, the macroscopic strain $\mathbf{E}$ and stress $\mathbf{\Sigma}$ are the averages[1] of the respec-tive microscopic quantities $\varepsilon(\mathbf{y})$ and $\sigma(\mathbf{y})$,

$$\mathbf{E} = \{\varepsilon\}, \qquad \mathbf{\Sigma} = \{\sigma\}, \qquad \{\cdot\} \equiv \frac{1}{|V|}\int_V (\cdot)\,\mathrm{d}V, \qquad (2.8)$$

where $|V| = \int_V \mathrm{d}V$.

The *localization* problem, i.e. the inverse of the homogenization procedure, is to determine the microscopic quantities, such as $\varepsilon(\mathbf{y})$ and $\sigma(\mathbf{y})$, knowing the macroscopic ones $\mathbf{E}$ or $\mathbf{\Sigma}$. The corresponding boundary value problem to be solved on the r.v.e. is the following

---

[1] In a more general setting, valid also in case of cracked or granular media, the macroscopic strain and stress are defined in terms of the boundary data

$$\mathbf{E} = \frac{1}{|V|}\int_{\partial V} \frac{1}{2}(\mathbf{n}\otimes\mathbf{u} + \mathbf{u}\otimes\mathbf{n})\,\mathrm{d}S, \qquad \mathbf{\Sigma} = \frac{1}{|V|}\int_{\partial V} \frac{1}{2}(\mathbf{t}\otimes\mathbf{y} + \mathbf{y}\otimes\mathbf{t})\,\mathrm{d}S,$$

where $\mathbf{u}$ is the displacement, $\mathbf{t}$ the surface traction, and $\mathbf{n}$ the unit outward normal. Expressions (2.8) are recovered for a coherent material in equilibrium (i.e. assuming continuity of displacements and mechanical equilibrium) by applying the divergence theorem.

$$\begin{cases} \text{div}\,\boldsymbol{\sigma} = \mathbf{0} & \text{(micro equilibrium)} \\ \{\varepsilon\} = \mathbf{E} \quad \text{or} \quad \{\boldsymbol{\sigma}\} = \boldsymbol{\Sigma} \\ \text{microscopic constitutive law} \\ \text{boundary conditions} \end{cases} \quad (2.9)$$

with $\mathbf{E}$ or $\boldsymbol{\Sigma}$ being the data. Three types of boundary conditions are classically applied for the above localization problem:

 i. linear displacement on the boundary,

$$\mathbf{u} = \mathbf{Ey} \quad \text{on} \quad \partial V; \quad (2.10)$$

 ii. uniform traction on the boundary,

$$\boldsymbol{\sigma}\mathbf{n} = \boldsymbol{\Sigma}\mathbf{n} \quad \text{on} \quad \partial V, \quad (2.11)$$

where $\mathbf{n}$ is the unit outward normal;
 iii. periodicity condition,

$$\tilde{\mathbf{u}}(\mathbf{y}^{+}) = \tilde{\mathbf{u}}(\mathbf{y}^{-}) \quad \text{and} \quad \boldsymbol{\sigma}(\mathbf{y}^{+})\mathbf{n}^{+} = -\boldsymbol{\sigma}(\mathbf{y}^{-})\mathbf{n}^{-}, \quad (2.12)$$

where $\tilde{\mathbf{u}}$ is the displacement fluctuation defined up to a rigid displacement, so that

$$\mathbf{u} = \mathbf{Ey} + \tilde{\mathbf{u}}, \quad (2.13)$$

and the boundary ∂V is decomposed into two parts $\partial V = \partial V^{-} \cup \partial V^{+}$, such that $\mathbf{n}^{+} = -\mathbf{n}^{-}$ at two associated points $\mathbf{y}^{+} \in \partial V^{+}$ and $\mathbf{y}^{-} \in \partial V^{-}$.

These three types of boundary conditions, illustrated in Fig. 2.1, are not equivalent unless the r.v.e. is sufficiently large so that the influence of the surface layer affected by the boundary conditions is negligible, cf. Hill [45].

It can be shown that a very important property, often called the *Hill's lemma*, cf. Hill [45], holds for a strain field ε, derived from an admissible displacement field, and for an admissible stress field $\boldsymbol{\sigma}$, viz.

$$\boxed{\{\boldsymbol{\sigma} \cdot \varepsilon\} = \boldsymbol{\Sigma} \cdot \mathbf{E},} \quad (2.14)$$

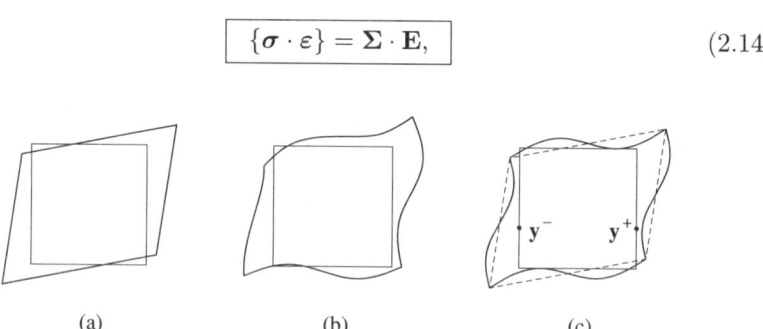

(a)   (b)   (c)

Fig. 2.1. Deformation of a r.v.e. for different boundary conditions: (a) linear displacement on ∂V, (b) uniform traction on ∂V, (c) periodic displacement fluctuations.

where a displacement field and a stress field are called admissible when they satisfy one set of boundary conditions (2.10) or (2.11), or (2.12) and, additionally, the stress field satisfies the micro equilibrium $(2.9)_1$.

Consider first the homogenization problem for heterogenous elasticity, i.e. the situation when the elastic stiffness and compliance tensors $\mathbf{L}$ and $\mathbf{M}$, respectively, depend on the position within the r.v.e., so that the constitutive law is

$$\boldsymbol{\sigma}(\mathbf{y}) = \mathbf{L}(\mathbf{y})\boldsymbol{\varepsilon}(\mathbf{y}), \qquad \boldsymbol{\varepsilon}(\mathbf{y}) = \mathbf{M}(\mathbf{y})\boldsymbol{\sigma}(\mathbf{y}), \qquad \mathbf{M} = \mathbf{L}^{-1}. \tag{2.15}$$

Under the usual assumptions of positive definiteness of $\mathbf{L}$ and $\mathbf{M}$, the solution of the localization problem (2.9) can be written in the following form

$$\boldsymbol{\varepsilon}(\mathbf{y}) = \mathbf{A}(\mathbf{y})\mathbf{E}, \qquad \boldsymbol{\sigma}(\mathbf{y}) = \mathbf{B}(\mathbf{y})\boldsymbol{\Sigma}, \tag{2.16}$$

depending on whether $\mathbf{E}$ or $\boldsymbol{\Sigma}$ is given. Here, the fourth-rank tensor $\mathbf{A}$ is the *strain concentration tensor*, and $\mathbf{B}$ is the *stress concentration tensor*. The concentration tensors are obtained by solving the respective boundary value problems on the r.v.e.[2]

The homogenization can now be performed by averaging the constitutive relations (2.15) using the solution of the localization problem (2.16). As a result, the *effective (overall) elastic moduli* tensors $\tilde{\mathbf{L}}$ and $\tilde{\mathbf{M}}$, relating the macroscopic stress and strain,

$$\boldsymbol{\Sigma} = \tilde{\mathbf{L}}\mathbf{E} \quad \text{(given } \mathbf{E}\text{)}, \qquad \mathbf{E} = \tilde{\mathbf{M}}\boldsymbol{\Sigma} \quad \text{(given } \boldsymbol{\Sigma}\text{)}, \tag{2.17}$$

are found in the form

$$\tilde{\mathbf{L}} = \{\mathbf{L}\mathbf{A}\} \quad \text{(given } \mathbf{E}\text{)}, \qquad \tilde{\mathbf{M}} = \{\mathbf{M}\mathbf{B}\} \quad \text{(given } \boldsymbol{\Sigma}\text{)}. \tag{2.18}$$

With account of the Hill's lemma (2.14), the symmetry of $\tilde{\mathbf{L}}$ and $\tilde{\mathbf{M}}$ can easily be verified by considering the macroscopic elastic strain energy density, namely

$$W = \{w\} = \frac{1}{2}\{\boldsymbol{\varepsilon} \cdot \mathbf{L}\boldsymbol{\varepsilon}\} = \frac{1}{2}\mathbf{E} \cdot \{\mathbf{A}^{\mathrm{T}} \mathbf{L}\mathbf{A}\}\mathbf{E} = \frac{1}{2}\mathbf{E} \cdot \tilde{\mathbf{L}}\mathbf{E}, \tag{2.19}$$

and similarly for the complementary energy $\frac{1}{2}\boldsymbol{\sigma} \cdot \mathbf{M}\boldsymbol{\sigma}$, so that we have

$$\boxed{\tilde{\mathbf{L}} = \{\mathbf{L}\mathbf{A}\} = \{\mathbf{A}^{\mathrm{T}} \mathbf{L}\mathbf{A}\}, \qquad \tilde{\mathbf{M}} = \{\mathbf{M}\mathbf{B}\} = \{\mathbf{B}^{\mathrm{T}}\mathbf{M}\mathbf{B}\}.} \tag{2.20}$$

It can be shown that the effective elastic moduli tensors $\tilde{\mathbf{L}}$ and $\tilde{\mathbf{M}}$, obtained for the same set of boundary conditions (2.10) or (2.11), or (2.12), are

---

[2] Often, the strain concentration tensor $\mathbf{A}$ is associated with boundary conditions (2.10) or (2.11), and the stress concentration tensor $\mathbf{B}$ is associated with (2.12). This is because the macroscopic strain $\mathbf{E}$ directly appears in (2.10) and (2.13), while the macroscopic stress $\boldsymbol{\Sigma}$ appears in (2.11).

equivalent in a sense that $\tilde{\mathbf{L}} = \tilde{\mathbf{M}}^{-1}$, cf. Suquet [146]. However, the effective moduli associated with *different* boundary conditions are not identical, although the difference decreases with increasing size of the r.v.e. Moreover, the different effective stiffness tensors can be ordered in the following way

$$\mathbf{E} \cdot \tilde{\mathbf{L}}_\Sigma \mathbf{E} \leq \mathbf{E} \cdot \tilde{\mathbf{L}}_{\mathrm{per}} \mathbf{E} \leq \mathbf{E} \cdot \tilde{\mathbf{L}}_\mathbf{E} \mathbf{E}, \qquad (2.21)$$

where $\tilde{\mathbf{L}}_\mathbf{E}$, $\tilde{\mathbf{L}}_\Sigma$, and $\tilde{\mathbf{L}}_{\mathrm{per}}$ correspond to boundary conditions (2.10), (2.11), and (2.12), respectively.

At the end of this section, let us consider the case of a heterogeneous inelastic solid. The microscopic constitutive relation of elasticity with eigenstrain is

$$\varepsilon^e(\mathbf{y}) = \mathbf{M}(\mathbf{y})\sigma(\mathbf{y}), \qquad \varepsilon(\mathbf{y}) = \varepsilon^e(\mathbf{y}) + \varepsilon^t(\mathbf{y}), \qquad (2.22)$$

where the total strain ε is decomposed into elastic ε^e and inelastic ε^t parts. The eigenstrain ε^t can be, for instance, the plastic strain, however, its actual origin and the evolution law need not be specified for the present purposes.

It can be shown that the macroscopic constitutive relation is also that of elasticity with eigenstrain,

$$\mathbf{E} = \tilde{\mathbf{M}}\Sigma + \mathbf{E}^t, \qquad \mathbf{E}^t = \{\mathbf{B}^T \varepsilon^t\}, \qquad (2.23)$$

but the inelastic macro-strain $\mathbf{E}^t$ is not a simple average of the inelastic micro-strains. In (2.23), the effective elastic compliance tensor $\tilde{\mathbf{M}}$ is that of a heterogeneous elastic material, cf. (2.20).

Let us decompose the micro-stress σ into a self-equilibrated residual stress σ^r and the remaining part $\mathbf{B}\Sigma$ that would occur if the material were elastic,

$$\sigma(\mathbf{y}) = \mathbf{B}(\mathbf{y})\Sigma + \sigma^r(\mathbf{y}). \qquad (2.24)$$

The macroscopic elastic strain energy density,

$$W = \{w\} = \frac{1}{2}\{(\varepsilon - \varepsilon^t) \cdot \mathbf{L}(\varepsilon - \varepsilon)^t\} = \frac{1}{2}\{\sigma \cdot \mathbf{M}\sigma\}, \qquad (2.25)$$

can now be written as a sum of the energy of the elastic macro-strains $\frac{1}{2}\Sigma \cdot \tilde{\mathbf{M}}\Sigma$ and the stored elastic energy associated with the residual stresses $\frac{1}{2}\{\sigma^r \cdot \mathbf{M}\sigma^r\}$, namely

$$W = \frac{1}{2}\Sigma \cdot \tilde{\mathbf{M}}\Sigma + \frac{1}{2}\{\sigma^r \cdot \mathbf{M}\sigma^r\}. \qquad (2.26)$$

2.4 Compatibility Conditions at a Bonded Interface

Consider an interface that separates two phases denoted by '+' and '−'. The material properties may be discontinuous across the interface, however, perfect bonding is assumed, so that the displacement field is continuous. As a result, the stress and the strain may suffer discontinuity at the interface.

The assumption of *continuity of displacements* implies the well-known geometrical compatibility condition restricting the jump of strain $\Delta\varepsilon$ to be in the form of the symmetrized diadic product of the interface normal $\mathbf{n}$ and a vector $\mathbf{c}$, namely

$$\Delta\varepsilon = \frac{1}{2}(\mathbf{c}\otimes\mathbf{n}+\mathbf{n}\otimes\mathbf{c}), \tag{2.27}$$

where

$$\Delta(\cdot) = (\cdot)^- - (\cdot)^+. \tag{2.28}$$

Furthermore, the assumption of *mechanical equilibrium* implies continuity of the normal traction vector, namely

$$\Delta\sigma\mathbf{n} = \mathbf{0}. \tag{2.29}$$

In the case of interface moving from '$-$' to '$+$' side with a normal speed $v_n > 0$, $\Delta(\cdot)$ is the *forward jump with respect to time*, cf. Petryk [102], being the usual spacial jump with a minus sign, $\Delta(\cdot) = -[\cdot]$. In that case, vector $\mathbf{c}$ in (2.27) has a clear physical interpretation, namely it is related to the velocity jump by $v_n\mathbf{c} = [\mathbf{v}]$.

Using the interior–exterior decomposition (2.1), the compatibility conditions (2.27) and (2.29) can be rewritten in the form

$$\Delta\varepsilon_{\mathrm{P}} = \mathbf{0}, \qquad \Delta\sigma_{\mathrm{A}} = \mathbf{0}. \tag{2.30}$$

Now, accounting for the property (2.5), it follows that

$$\Delta\sigma \cdot \Delta\varepsilon = 0. \tag{2.31}$$

2.5 Interfacial Relationships

Assume that the constitutive equations of linear isothermal anisotropic elasticity with eigenstrain hold for the materials at both sides of the discontinuity surface,

$$\varepsilon^\pm = \mathbf{M}^\pm\sigma^\pm + \varepsilon^{\mathrm{t}\pm}, \qquad \sigma^\pm = \mathbf{L}^\pm(\varepsilon^\pm - \varepsilon^{\mathrm{t}\pm}), \tag{2.32}$$

where $\mathbf{M}^\pm$ and $\mathbf{L}^\pm = (\mathbf{M}^\pm)^{-1}$ are the elastic moduli tensors that possess usual symmetries and are positive definite. The eigenstrain $\varepsilon^{\mathrm{t}\pm}$ may result from phase transformation, plastic deformation, thermal strain, etc., but here its origin and evolution law (if any) may be left unspecified.

Assuming that the state in the '$+$' phase is known, the compatibility conditions (2.27) and (2.29) provide six equations sufficient to determine the state in the '$-$' phase. As a result, the stress and strain in the '$-$' phase are expressed in terms of the '$+$' values by the following interfacial relationships

$$\Delta\varepsilon = -\mathbf{P}^0(\Delta\mathbf{L}\varepsilon^+ - \Delta\sigma^t),$$
$$\Delta\sigma = -\mathbf{S}^0(\Delta\mathbf{M}\sigma^+ + \Delta\varepsilon^t),$$

(2.33)

where $\sigma^{t\pm} = \mathbf{L}^\pm\varepsilon^{t\pm}$, and $\mathbf{P}^0$ and $\mathbf{S}^0$ are fourth-rank tensors that depend on the elastic properties of the '$-$' phase and on the orientation of the interface, specified by its normal $\mathbf{n}$.

A slightly different form of the interfacial relationships (2.33) and the coordinate-invariant expressions for the operators $\mathbf{P}^0$ and $\mathbf{S}^0$ can be found in Hill [47]. Explicit expressions for $\mathbf{P}^0$ and $\mathbf{S}^0$, expressed in the intrinsic coordinate system, are also provided in Appendix A.3.

2.6 Micro-Macro Transition for Simple Laminates

Consider now a two-phase (two-component) laminate with the volume fractions of two homogeneous '$-$' and '$+$' phases being, respectively, $\eta^- = \eta$ and $\eta^+ = (1 - \eta)$, where $0 \le \eta \le 1$, cf. Fig. 2.2. Such microstructure will be referred to as a *simple laminate*. The parallel interfaces separating the two constituents are characterized by the normal vector $\mathbf{n}$.

A peculiar property of laminates is that the stresses and strains are uniform within all layers of the same component, so that the interfacial relationships (2.33) hold not only locally at the interfaces but in the whole volume.[3] Spatial arrangement of the layers is thus fully irrelevant (a periodic layout may be chosen to fix the attention), and the microstructure is fully characterized by the volume fraction η. Similarly, the spatial variation of strains, stresses and other fields, i.e. their dependence on the position $\mathbf{y}$ within the r.v.e., is not relevant, and it is sufficient to distinguish the respective quantities within each phase by a superscript '$-$' or '$+$'.

In view of the compatibility conditions (2.27) and (2.29), relating now the strains and stresses within each phase, it can easily be verified that the Hill's lemma (2.14) holds automatically for the laminate fields. For this reason the

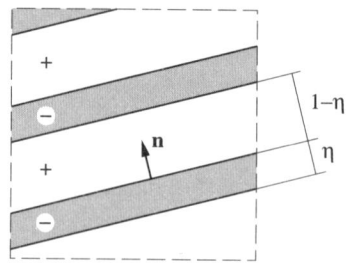

Fig. 2.2. Simple laminate.

---

[3] In the case of a multi-component laminate, the interfacial relationships (2.33) hold between any two layers, not necessarily the adjacent ones.

choice of the r.v.e. is quite arbitrary. To fix the attention, as the r.v.e., we adopt an oblique cylinder with an arbitrary base at $y_n = \mathbf{y} \cdot \mathbf{n} = \text{const}$, and of sufficient height.

Let us specify the free energy density function of each phase, per unit volume, consistent with the constitutive stress–strain relations (2.32) of anisotropic elasticity with eigenstrain, in the form

$$\phi^\pm = \phi_0^\pm + \frac{1}{2}(\varepsilon^\pm - \varepsilon^{t\pm}) \cdot \mathbf{L}^\pm(\varepsilon^\pm - \varepsilon^{t\pm}) = \phi_0^\pm + \frac{1}{2}\sigma^\pm \cdot \mathbf{M}^\pm\sigma^\pm, \quad (2.34)$$

where $\phi_0^\pm$ is the free energy density in the stress-free state.

Applying the averaging rules (2.8) and the compatibility conditions (2.27) and (2.29), the following macroscopic constitutive relation for the laminate is obtained,[4]

$$\mathbf{E} = \tilde{\mathbf{M}}\boldsymbol{\Sigma} + \mathbf{E}^t, \qquad \boldsymbol{\Sigma} = \tilde{\mathbf{L}}(\mathbf{E} - \mathbf{E}^t), \quad (2.35)$$

as a specification of the general expressions provided in Sect. 2.3, with

$$\mathbf{E}^t = \{\mathbf{B}^T \varepsilon^t\} = \eta\,(\mathbf{B}^-)^T \varepsilon^{t-} + (1-\eta)\,(\mathbf{B}^+)^T \varepsilon^{t+}. \quad (2.36)$$

Furthermore, the local stresses within the phases are found to be equal to

$$\sigma^- = \mathbf{B}^-\boldsymbol{\Sigma} + (1-\eta)\,\mathbf{S}(\varepsilon^{t+} - \varepsilon^{t-}),$$
$$\sigma^+ = \mathbf{B}^+\boldsymbol{\Sigma} + \eta\,\mathbf{S}(\varepsilon^{t-} - \varepsilon^{t+}), \quad (2.37)$$

and the dual expressions for the local strains are

$$\varepsilon^- = \mathbf{A}^-\,\mathbf{E} - (1-\eta)\,\mathbf{P}(\sigma^{t+} - \sigma^{t-}),$$
$$\varepsilon^+ = \mathbf{A}^+\,\mathbf{E} - \eta\,\mathbf{P}(\sigma^{t-} - \sigma^{t+}). \quad (2.38)$$

Here $\mathbf{A}^\pm$ and $\mathbf{B}^\pm$ are, respectively, the strain and stress concentration tensors (constant within the layers), and $\mathbf{P}$ and $\mathbf{S}$ are fourth-rank tensors. Finally, the macroscopic free energy is given by

$$\Phi = \{\phi\} = \Phi_0 + \frac{1}{2}\,(\mathbf{E} - \mathbf{E}^t) \cdot \tilde{\mathbf{L}}(\mathbf{E} - \mathbf{E}^t) = \Phi_0 + \frac{1}{2}\boldsymbol{\Sigma} \cdot \tilde{\mathbf{M}}\boldsymbol{\Sigma}, \quad (2.39)$$

with

$$\Phi_0 = \eta\phi_0^- + (1-\eta)\phi_0^+ + \frac{1}{2}\eta(1-\eta)(\varepsilon^{t+} - \varepsilon^{t-}) \cdot \mathbf{S}(\varepsilon^{t+} - \varepsilon^{t-}), \quad (2.40)$$

where the last term on right-hand side of (2.40) is recognized to be the elastic energy associated with the residual stresses, cf. (2.26).

The fourth-rank tensors $\tilde{\mathbf{M}}$, $\tilde{\mathbf{L}}$, $\mathbf{A}^\pm$, $\mathbf{B}^\pm$, $\mathbf{P}$, and $\mathbf{S}$ depend on the elasticity tensors $\mathbf{M}^\pm$ and $\mathbf{L}^\pm$, the volume fraction η, and the interface normal $\mathbf{n}$. The

---

[4] As the strains and stresses in the laminate are piecewise constant, solution of the localization problem (2.9) is straightforward and amounts to solution of a system of linear algebraic equations.

respective analytic formulae in the matrix form are given in Appendix A.4. Note that it is not easy to find in the literature a reference with a complete set of those formulae, although the approach to derive them is rather standard. For instance, the expressions for the effective elastic stiffness tensor in the case of arbitrary anisotropy of the layers can be found in Gałka et al. [31], and an expression analogous to (2.40), but in a different notation, was given by Roytburd [119].

Remark 2.1. For $\eta = 0$, the operators $\mathbf{P}$ and $\mathbf{S}$ in (2.37) and (2.38) reduce to the respective operators $\mathbf{P}^0$ and $\mathbf{S}^0$ which appear in the interfacial relationships (2.33), see Appendix A.

Remark 2.2. The compatibility conditions (2.30) imply that the interior part of the local strain and the exterior part of the local stress are continuous and equal to the respective parts of the macroscopic variables, thus

$$\mathbf{E}_\mathrm{P} = \{\varepsilon_\mathrm{P}\} = \varepsilon_\mathrm{P}^+ = \varepsilon_\mathrm{P}^-, \qquad \mathbf{\Sigma}_\mathrm{A} = \{\sigma_\mathrm{A}\} = \sigma_\mathrm{A}^+ = \sigma_\mathrm{A}^-. \qquad (2.41)$$

This property is the basis of an alternative method of computing the effective moduli of laminated microstructures. As shown by El Omri et al. [27], homogenization of laminates of arbitrary constitutive behaviour can conveniently be carried out using the constitutive equations in the mixed form, cf. Sect. 3.2. In fact, having the local constitutive equations in the mixed form, the mixed form of the macroscopic constitutive equations is obtained by simple averaging of the local ones.

3

Homogeneous Surface Layers

Abstract: Mixed form of constitutive relations of thin homogeneous layers is derived for elastic, elastic-plastic, and rigid-plastic material models. This formalism is then applied to derive a phenomenological model of real contact evolution in metal forming processes. The predictions of the model are compared to other models and to available experimental data. Remarks are also provided regarding the application of the derived evolution law to modelling of friction in metal forming processes.

3.1 Motivation and Preliminaries

The main part of this chapter is devoted to the phenomenological modelling of evolution of real contact area in metal forming processes. The model of Stupkiewicz and Mróz [138] is presented in Sect. 3.3 and, following Stupkiewicz and Mróz [136, 137], its application to modelling of friction in metal forming processes is discussed in Sect. 3.3.7. In the present phenomenological modelling, workpiece asperities and a thin subsurface layer of inhomogeneous deformation induced by their flattening (the boundary layer in the terminology of Chaps. 4 and 5) is assumed to be represented by an equivalent homogeneous layer. A constitutive law is then postulated for the surface layer with the aim of describing the effect of macroscopic (bulk) plastic deformation on asperity flattening and the related evolution of the real contact area.

However, thin, homogeneous layers are met in the engineering practice also in other applications such as coatings, epitaxial layers, etc. Therefore, in Sect. 3.2 below, the thin layers are first discussed from a broader perspective.

The analysis is restricted to the case of infinitesimally thin layers, so that the stress and strain in the substrate are not affected by the presence of the layer, while the deformation of the layer is fully determined by the deformation of the substrate. This also implies that the thickness of the layer is negligible compared to the radius of curvature of the surface.

In view of the thin layer assumption and in view of homogeneity of the layer, the stresses and the strains are constant across the layer. Perfect bonding of the layer to the substrate is also assumed, so that the compatibility conditions (2.27) and (2.29) are satisfied at the surface layer-substrate interface. Denoting by σ and ε the stress and strain in the surface layer and by Σ and E the stress and strain in the substrate in the vicinity of the surface layer-substrate interface, the compatibility conditions (2.27) and (2.29) can be written as

$$\varepsilon_P = E_P, \qquad \sigma_A = \Sigma_A, \qquad (3.1)$$

where the subscripts $(\cdot)_P$ and $(\cdot)_A$ denote, respectively, the interior and exterior parts, cf. the interior–exterior decomposition, Sect. 2.2.

As shown below, using the compatibility conditions (3.1) combined with the constitutive relations of the layer, the stress and strain within the layer can be fully determined in terms of E_P and Σ_A, the latter being directly related to the contact traction, cf. Sect. 2.2. This leads to the *mixed form*[1] of constitutive relations, cf. El Omri et al. [27].

Below, the constitutive equations in the mixed form are provided for elastic, elastic-plastic, and rigid-plastic layers. Special attention is paid to the rigid-plastic case as a background for the phenomenological model discussed in Sect. 3.3.

As the interior–exterior decomposition is an essential tool of the present analysis, the Kelvin matrix notation, which allows a convenient application of this decomposition in the intrinsic coordinate system, cf. Appendix A, is used throughout this chapter.

3.2 Mixed Form of Constitutive Equations

3.2.1 Elastic Layer

Consider first an elastic layer, and denote by ε^t the eigenstrain, so that the stress-free configuration of the layer does not correspond to a zero total strain. The corresponding constitutive equation, written using the matrix notation in the intrinsic Cartesian coordinate system, cf. equation (A.7), takes the form

$$\left\{ \begin{array}{c} \sigma_P \\ \sigma_A \end{array} \right\} = \left[\begin{array}{cc} L_{PP} & L_{PA} \\ L_{AP} & L_{AA} \end{array} \right] \left\{ \begin{array}{c} \varepsilon_P - \varepsilon_P^t \\ \varepsilon_A - \varepsilon_A^t \end{array} \right\}. \qquad (3.2)$$

The partial inversion of (3.2) gives the following mixed form of the constitutive equation

$$\left\{ \begin{array}{c} \sigma_P \\ \varepsilon_A - \varepsilon_A^t \end{array} \right\} = \left[\begin{array}{cc} L_{PP} - L_{PA}L_{AA}^{-1}L_{AP} & L_{PA}L_{AA}^{-1} \\ -L_{AA}^{-1}L_{AP} & L_{AA}^{-1} \end{array} \right] \left\{ \begin{array}{c} E_P - \varepsilon_P^t \\ \Sigma_A \end{array} \right\}, \qquad (3.3)$$

---

[1] Or the hybrid form in the terminology of El Omri et al. [27].

where, in view of the compatibility conditions (3.1), ε_P and σ_A have been replaced by the respective components of the substrate strain and stress. Positive definiteness of $\mathbf{L}$ guarantees that $\mathbf{L_{AA}}$ is also positive definite, so that $\mathbf{L_{AA}}$ can be inverted.

3.2.2 Elastic-Plastic Layer

Consider now an elastic-plastic layer. The total strain rate $\dot{\varepsilon}$ is decomposed into elastic part $\dot{\varepsilon}^e$ and plastic part $\dot{\varepsilon}^p$, so that the rate of stress is given by

$$\dot{\sigma} = \mathbf{L}\dot{\varepsilon}^e = \mathbf{L}(\dot{\varepsilon} - \dot{\varepsilon}^p). \tag{3.4}$$

The yield condition of Huber-von Mises plasticity with isotropic hardening is

$$F(\sigma, \varepsilon^p) = \sqrt{\frac{3}{2} \sigma \cdot \mathbf{\Pi}^d \sigma} - \sigma_y(\varepsilon^p) \leq 0, \tag{3.5}$$

where σ_y is the yield stress, given as a function of the equivalent plastic strain ε^p, and $\mathbf{\Pi}^d$ is the projection matrix onto the deviatoric space,

$$\mathbf{\Pi}^d = \frac{1}{3} \begin{bmatrix} 2 & -1 & -1 & 0 & 0 & 0 \\ & 2 & -1 & 0 & 0 & 0 \\ & & 2 & 0 & 0 & 0 \\ & & & 3 & 0 & 0 \\ & & & & 3 & 0 \\ symmetric & & & & & 3 \end{bmatrix}, \tag{3.6}$$

given here in the form corresponding to the arrangement of the stress vector components as in equation (A.1). The plastic strain rate $\dot{\varepsilon}^p$ is given by the associated flow rule

$$\dot{\varepsilon}^p = \gamma\,\mu, \qquad \mu = \frac{\partial F}{\partial \sigma} = \frac{3}{2\sigma_y} \mathbf{\Pi}^d \sigma, \qquad \gamma \geq 0, \tag{3.7}$$

where γ is the plastic multiplier, and the equivalent plastic strain ε^p is defined by

$$\dot{\varepsilon}^p = \gamma = \sqrt{\frac{2}{3} \dot{\varepsilon}^p \cdot \dot{\varepsilon}^p}. \tag{3.8}$$

Finally, the plastic flow with $\gamma > 0$ is only possible if $F = 0$, and this is expressed in the form of the complementarity condition,

$$\gamma F = 0. \tag{3.9}$$

Following the classical argument of the plasticity theory, e.g. Hill [44], Simo and Hughes [127], the constitutive rate-equations are given by

$$\dot{\sigma} = \mathbf{L}^t \dot{\varepsilon}, \tag{3.10}$$

where $\mathbf{L}^t$ is the matrix of tangent moduli: $\mathbf{L}^t = \mathbf{L}$ if the state is elastic, and $\mathbf{L}^t = \mathbf{L}^{ep}$ if the state is plastic. Here, $\mathbf{L}^{ep}$ is the matrix of elastoplastic tangent moduli,

$$\mathbf{L}^{ep} = \mathbf{L} - \frac{1}{g}\boldsymbol{\lambda} \otimes \boldsymbol{\lambda}, \qquad g = \sigma_y' + \boldsymbol{\mu} \cdot \mathbf{L}\boldsymbol{\mu}, \qquad \boldsymbol{\lambda} = \mathbf{L}\boldsymbol{\mu}. \qquad (3.11)$$

Similarly to the elastic case, after the interior and exterior components of stress and strain rates are introduced, the mixed form of the constitutive equation (3.10) is obtained by partial inversion, viz.

$$\left\{ \begin{matrix} \dot{\boldsymbol{\sigma}}_P \\ \dot{\boldsymbol{\varepsilon}}_A \end{matrix} \right\} = \left[\begin{matrix} \mathbf{L}_{PP}^t - \mathbf{L}_{PA}^t(\mathbf{L}_{AA}^t)^{-1}\mathbf{L}_{AP}^t & \mathbf{L}_{PA}^t(\mathbf{L}_{AA}^t)^{-1} \\ -(\mathbf{L}_{AA}^t)^{-1}\mathbf{L}_{AP}^t & (\mathbf{L}_{AA}^t)^{-1} \end{matrix} \right] \left\{ \begin{matrix} \dot{\mathbf{E}}_P \\ \dot{\boldsymbol{\Sigma}}_A \end{matrix} \right\}. \qquad (3.12)$$

Clearly, the above partial inversion is only possible if sub-matrix $\mathbf{L}_{AA}^t$ is not singular, $\det \mathbf{L}_{AA}^t \neq 0$. This is guaranteed by simple ellipticity of $\mathbf{L}^t$. Moreover, if $\mathbf{L}^t$ is strongly elliptic, then $\mathbf{L}_{AA}^t$ is positive definite, cf. Hill [47]. Note that $\mathbf{L}_{AA}^t$ is, essentially, a representation of the acoustic tensor $\mathbf{n}\mathbf{L}^t\mathbf{n}$ with the rows and columns corresponding to shear components multiplied by $\sqrt{2}$. It thus inherits many properties of the acoustic tensor.

3.2.3 Rigid-Plastic Layer

Consider finally the case of a rigid-perfectly plastic layer. The yield condition and the flow rule of the associative Huber-von Mises plasticity are specified by

$$F(\boldsymbol{\sigma}) = \sqrt{\frac{3}{2}\boldsymbol{\sigma} \cdot \boldsymbol{\Pi}^d\boldsymbol{\sigma}} - \sigma_y \leq 0, \quad \dot{\boldsymbol{\varepsilon}} = \gamma \frac{\partial F}{\partial \boldsymbol{\sigma}} = \frac{3\gamma}{2\sigma_y}\boldsymbol{\Pi}^d\boldsymbol{\sigma}, \quad \gamma \geq 0, \quad (3.13)$$

with the complementarity condition (3.9). Below, the case of plastic flow with $\gamma > 0$ is only considered.

After the flow rule $(3.13)_2$ is expressed in terms of interior and exterior components, namely

$$\left\{ \begin{matrix} \dot{\boldsymbol{\varepsilon}}_P \\ \dot{\boldsymbol{\varepsilon}}_A \end{matrix} \right\} = \frac{3\gamma}{2\sigma_y} \left[\begin{matrix} \boldsymbol{\Pi}_{PP}^d & \boldsymbol{\Pi}_{PA}^d \\ \boldsymbol{\Pi}_{AP}^d & \boldsymbol{\Pi}_{AA}^d \end{matrix} \right] \left\{ \begin{matrix} \boldsymbol{\sigma}_P \\ \boldsymbol{\sigma}_A \end{matrix} \right\}, \qquad (3.14)$$

it can be partially inverted to yield

$$\left\{ \begin{matrix} \boldsymbol{\sigma}_P \\ \dot{\boldsymbol{\varepsilon}}_A \end{matrix} \right\} = \left[\begin{matrix} \frac{2\sigma_y}{3\gamma}(\boldsymbol{\Pi}_{PP}^d)^{-1} & -(\boldsymbol{\Pi}_{PP}^d)^{-1}\boldsymbol{\Pi}_{PA}^d \\ \boldsymbol{\Pi}_{AP}^d(\boldsymbol{\Pi}_{PP}^d)^{-1} & \frac{3\gamma}{2\sigma_y}[\boldsymbol{\Pi}_{AA}^d - \boldsymbol{\Pi}_{AP}^d(\boldsymbol{\Pi}_{PP}^d)^{-1}\boldsymbol{\Pi}_{PA}^d] \end{matrix} \right] \left\{ \begin{matrix} \dot{\mathbf{E}}_P \\ \dot{\boldsymbol{\Sigma}}_A \end{matrix} \right\}, \qquad (3.15)$$

where, additionally, the compatibility conditions $\boldsymbol{\sigma}_A = \boldsymbol{\Sigma}_A$ and $\dot{\boldsymbol{\varepsilon}}_P = \dot{\mathbf{E}}_P$ have been used. Note that, although the projection operator $\boldsymbol{\Pi}^d$ is singular, the sub-matrix $\boldsymbol{\Pi}_{PP}^d$ is not singular and thus it can be inverted.

In order to express the plastic multiplier γ in terms of the control variables $\dot{\mathbf{E}}_P$ and $\boldsymbol{\Sigma}_A$, the equivalent strain rate $\dot{\varepsilon}$, defined as

$$\dot{\varepsilon} = \gamma = \sqrt{\frac{2}{3}\, \dot{\boldsymbol{\varepsilon}} \cdot \dot{\boldsymbol{\varepsilon}}}, \tag{3.16}$$

is split into interior and exterior contributions, namely

$$\dot{\varepsilon} = \sqrt{(\dot{\varepsilon}_P)^2 + (\dot{\varepsilon}_A)^2}, \qquad \dot{\varepsilon}_P = \sqrt{\frac{2}{3}\, \dot{\boldsymbol{\varepsilon}}_P \cdot \dot{\boldsymbol{\varepsilon}}_P}, \qquad \dot{\varepsilon}_A = \sqrt{\frac{2}{3}\, \dot{\boldsymbol{\varepsilon}}_A \cdot \dot{\boldsymbol{\varepsilon}}_A}. \tag{3.17}$$

Adopting a coordinate system such that x_1- and x_2-axes are the principal directions of the interior strain rate component $\dot{\boldsymbol{\varepsilon}}_P$ (so that $\dot{\varepsilon}_{12} = 0$), an angle ϕ can be introduced, such that

$$\tan \phi = \frac{\dot{\varepsilon}_{22}}{\dot{\varepsilon}_{11}}, \qquad \dot{\boldsymbol{\varepsilon}}_P = \sqrt{\frac{3}{2}}\, \dot{\varepsilon}_P \{\cos \phi, \sin \phi, 0\}. \tag{3.18}$$

For instance, $\phi = 0$ in the plane strain conditions, $\varepsilon_{22} = 0$.

Now, the following relation between the equivalent total strain rate $\dot{\varepsilon}$ and its interior component $\dot{\varepsilon}_P$ follows from the flow rule $(3.13)_2$,

$$\dot{\varepsilon} = \dot{\varepsilon}_P \sqrt{\frac{2 + \sin 2\phi}{1 - (T_T/k)^2}}, \tag{3.19}$$

where $T_T = \sqrt{\Sigma_{13}^2 + \Sigma_{23}^2}$ is the norm of the tangential component of surface traction vector (i.e. the friction stress if the contact surface layer is considered), and $k = \sigma_y/\sqrt{3}$ is the yield stress in shear.

Equation (3.19) allows expressing the plastic multiplier $\gamma = \dot{\varepsilon}$ in terms of $\dot{\mathbf{E}}_P$ and $\boldsymbol{\Sigma}_A$, so that $\boldsymbol{\sigma}_P$ and $\dot{\varepsilon}_A$ are expressed by (3.15) solely in terms of $\dot{\mathbf{E}}_P$ and $\boldsymbol{\Sigma}_A$. The mixed form of the yield condition can finally be derived by combining $(3.13)_1$ and (3.15), namely

$$F^*(\boldsymbol{\Sigma}_A, \dot{\mathbf{E}}_P) = \sqrt{\frac{3}{2}\, \boldsymbol{\Sigma}_A \cdot \boldsymbol{\Pi}^*_{AA} \boldsymbol{\Sigma}_A + \frac{2\sigma_y^2}{3\gamma^2}\, \dot{\mathbf{E}}_P \cdot (\boldsymbol{\Pi}^d_{PP})^{-1} \dot{\mathbf{E}}_P} - \sigma_y \leq 0, \tag{3.20}$$

where

$$\boldsymbol{\Pi}^*_{AA} = \boldsymbol{\Pi}^d_{AA} - \boldsymbol{\Pi}^d_{AP}(\boldsymbol{\Pi}^d_{PP})^{-1}\boldsymbol{\Pi}^d_{PA} = \begin{bmatrix} 0 & 0 & 0 \\ 0 & 1 & 0 \\ 0 & 0 & 1 \end{bmatrix}, \tag{3.21}$$

and

$$\gamma = \sqrt{\frac{2}{3}\, \dot{\mathbf{E}}_P \cdot \dot{\mathbf{E}}_P\, \frac{2 + \sin 2\phi}{1 - (T_T/k)^2}}. \tag{3.22}$$

It follows from (3.21) that the normal contact traction component $T_N = \Sigma_{33}$ does not directly affect the yield condition in the mixed form (note the zeros at the corresponding positions of $\boldsymbol{\Pi}^*_{AA}$).

3.3 Phenomenological Model of Evolution of Real Contact Area in Metal Forming

3.3.1 Real Contact Area and Friction in Metal Forming

Contact phenomena play a fundamental role in metal forming processes since the frictional contact interactions between the workpiece and the tooling, in fact, control the deformation of the workpiece. Moreover, surface finish of the product, service life of tools, impact of toxic lubricants on the environment, etc., are important economic issues related to contact phenomena. Therefore, accurate modelling of these phenomena is essential for reliable and accurate simulations of forming processes.

Contact of rough bodies occurs at small spots, called real contacts, which usually constitute a small fraction of the nominal contact area. This strongly affects the contact phenomena, such as friction and heat flow between bodies in contact. For instance, in simple models of friction, the macroscopic friction stress is determined by the shear resistance of adhesive junctions and by the real contact area fraction, cf. Bowden and Tabor [16]. The real contact area fraction is also a fundamental parameter that affects the heat flow through the contact surface, cf. Cooper et al. [23].

At the micro-scale, the contact phenomena are governed by interaction of surface asperities. The two basic mechanisms of friction, i.e. adhesion and ploughing, can be associated with two basic asperity interaction mechanisms illustrated in Fig. 3.1. In metal forming processes, flattening of workpiece asperities is the main asperity interaction mechanism. A respective microme-chanical model of friction has been proposed by Wanheim and Bay [10, 156] who considered the adhesive friction mechanism and rigid-plastic flattening of workpiece asperities according to the rough workpiece-smooth tool (RW-ST) interaction mode, cf. Fig. 3.1(a). In that model, the friction stress is proportional to the real contact area which, in turn, is related to the normal contact pressure. Accordingly, at low contact pressures, the friction stress is proportional to the contact pressure, like in the Coulomb law, and, at high contact pressures, a threshold friction stress is obtained, as the real contact area fraction approaches then unity.

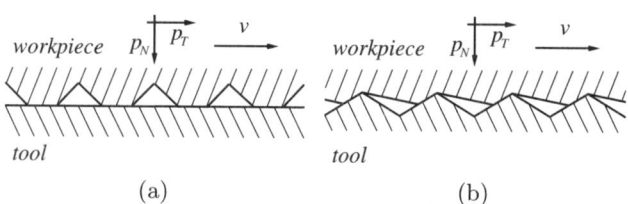

Fig. 3.1. Basic asperity interaction modes in metal forming: (a) rough workpiece-smooth tool; (b) smooth workpiece-rough tool.

The ploughing friction mechanism is associated with plastic deformations induced by hard asperities or abrasive particles which, upon relative motion of the surfaces, plough through the softer surface, cf. the smooth workpiece-rough tool (SW-RT) interaction mode, Fig. 3.1(b). Corresponding models have been developed using the slip line field technique or the upper bound method, cf. Challen and Oxley [21], Petryk [101], Avitzur and Nakamura [5], Azarkin and Richmond [6].

The two basic asperity interaction modes have been combined by Mróz and Stupkiewicz [84] by assuming separation of scales. In that model, the workpiece asperities are flattened according to the RW-ST model, and, at a lower scale, the tool asperities plough through the plateaus of flattened workpiece asperities. The model accounts for contact memory effects and transient states, and the real contact area fraction is a contact state variable representing the irreversible asperity flattening process. Additional contact state variables, such as accumulated friction work and sliding distance, have been included in the modelling by de Souza Neto et al. [25] and Gearing et al. [32], and hard particles of oxide layer have been accounted for by Stupkiewicz and Mróz [135].

The models discussed above neglect an important effect of macroscopic plastic deformations. In fact, severe plastic deformations of the workpiece constitute an essential and distinctive feature of contact conditions in metal forming operations. Both experiment and theory predict that asperities are flattened more easily if the underlying bulk material deforms plastically. Respective micro-mechanical models have been developed on the basis of the slip line method (Sutcliffe [147]), the upper bound approach (Wilson and Sheu [160, 162], Kimura and Childs [61]), and finite element solutions (Korzekwa et al. [69], Ike [51]). In these micro-mechanical models, idealized process conditions and contact geometries are assumed as required by the applied solution techniques.

A phenomenological model of real contact area evolution accounting for the effects of macroscopic plastic deformations has been developed by Stupkiewicz and Mróz [138]. In this model, a thin, homogeneous surface layer is considered which is assumed to represent the subsurface layer of inhomogeneous deformations induced by deforming asperities. Constitutive equations of this equivalent surface layer are developed in a phenomenological way, although a micromechanical reasoning is utilized to reduce the number of adjustable functions and parameters. Next, following in essence the scheme outlined in Sect. 3.2.3, the mixed form of the constitutive equations is derived, which provides a relation between the contact tractions, real contact area fraction, and macroscopic plastic strain. This model is outlined in the subsequent sections, and its application to modelling of friction is provided.

3.3.2 Basic Assumptions

The basic idea of the model is to replace workpiece asperities and a thin sub-surface layer of inhomogeneous plastic deformations by an equivalent homogeneous surface layer, cf. Fig. 3.2. Since the thickness of the layer is small, as compared to the characteristic dimensions of the workpiece, the stress σ and the strain ε are assumed constant across the layer. These quantities should be understood as phenomenological representations of averaged stresses and strains within the real surface layer (or boundary layer, in the terminology of Chaps. 4 and 5). By Σ and $\mathbf{E}$, we denote the macroscopic stress and strain at the points adjacent to the surface layer. In view of displacement continuity and stress equilibrium, the compatibility conditions (3.1) are assumed to hold at the surface layer-substrate interface.

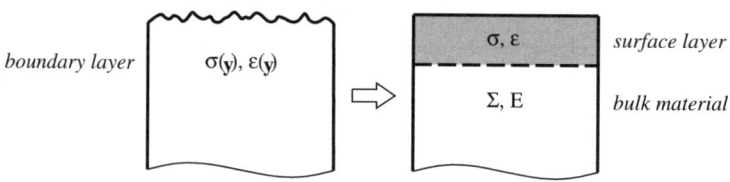

Fig. 3.2. Homogeneous surface layer representing workpiece asperities and a thin subsurface layer of inhomogeneous deformation.

In the present approach, asperity flattening and evolution of real contact area are modelled by postulating a constitutive law for the homogeneous surface layer. A central assumption is that the surface layer is *weakened*, with respect to the bulk material, due to the inhomogeneity of plastic deformations at the micro-scale. This in agreement with the experimental results of Sutcliffe [147], and also with the micromechanical predictions presented in Sect. 5.3.2.

The tool surface in contact with the workpiece is assumed smooth which corresponds to the RW-ST interaction mode, cf. Fig. 3.1(a). Accordingly, the real contact area fraction is used as a measure of the workpiece asperity flattening process. The real contact area fraction, denoted by α, is defined as the ratio of the total area A_r of real contacts to the nominal area A_n, $\alpha = A_r/A_n$.

As the model is specialized for metal forming processes, the elastic strains are neglected in the present modelling, and both the bulk material and the surface layer are assumed to be rigid-plastic and obeying a rate-independent constitutive law. Further, material properties and surface roughness are assumed to be isotropic, so that there are no privileged directions of plastic straining and friction slip. Behaviour of the bulk material is thus assumed to be governed by the constitutive equations (3.13) of a rigid-perfectly plastic material, expressed in terms of Σ and $\mathbf{E}$. Constitutive equations of the surface layer are specified below.

3.3.3 Asperity Flattening Condition

Weakening of the surface layer is directly related to the deformation induced by asperity flattening. The yield condition of the layer is thus assumed to additionally depend on the real contact area fraction α which is a measure of the flattening. The *asperity flattening condition*, i.e. the yield condition of the surface layer, is adopted in a form analogous to (3.13), namely

$$F^{\mathrm{l}}(\boldsymbol{\sigma}, \alpha) = \sqrt{\frac{3}{2}\,\boldsymbol{\sigma} \cdot \boldsymbol{\Pi}^{\mathrm{l}}(\alpha)\boldsymbol{\sigma}} - \sigma_{\mathrm{y}}^{\mathrm{l}}(\alpha) = 0, \tag{3.23}$$

where $\boldsymbol{\Pi}^{\mathrm{l}}$ and $\sigma_{\mathrm{y}}^{\mathrm{l}}$ are assumed to depend on α. Furthermore, an associated flow rule is assumed, thus

$$\dot{\boldsymbol{\varepsilon}} = \gamma^{\mathrm{l}} \frac{\partial F^{\mathrm{l}}}{\partial \boldsymbol{\sigma}} = \frac{3\gamma^{\mathrm{l}}}{2\sigma_{\mathrm{y}}^{\mathrm{l}}} \boldsymbol{\Pi}^{\mathrm{l}}\boldsymbol{\sigma}, \qquad \gamma^{\mathrm{l}} \geq 0. \tag{3.24}$$

There is some freedom in choosing a particular form of the operator $\boldsymbol{\Pi}^{\mathrm{l}}$ in (3.23), but clearly this choice is crucial for the predicting capabilities of the model. The following form of $\boldsymbol{\Pi}^{\mathrm{l}}$,

$$\boldsymbol{\Pi}^{\mathrm{l}} = \boldsymbol{\Pi}^{\mathrm{d}} + g_1(\alpha)\boldsymbol{\Pi}^{\mathrm{c}}, \qquad \boldsymbol{\Pi}^{\mathrm{c}} = \mathrm{diag}[0, 0, \tfrac{2}{3}, 0, 1, 1], \tag{3.25}$$

proved to provide satisfactory results. Here, $\boldsymbol{\Pi}^{\mathrm{d}}$ is the projection matrix onto the deviatoric space, cf. (3.6), $\boldsymbol{\Pi}^{\mathrm{c}}$ is a diagonal matrix, and the non-zero terms in $\boldsymbol{\Pi}^{\mathrm{c}}$ are exactly the diagonal terms of $\boldsymbol{\Pi}^{\mathrm{d}}$ which correspond to the exterior part of the stress, i.e. to the components of the contact traction vector. Furthermore, let $\sigma_{\mathrm{y}}^{\mathrm{l}}(\alpha) = g_2(\alpha)\sigma_{\mathrm{y}}$, where σ_{y} is the yield stress of the bulk material, so that the yield condition of the surface layer becomes

$$\boxed{F^{\mathrm{l}}(\boldsymbol{\sigma}, \alpha) = \sqrt{\frac{3}{2}\,\boldsymbol{\sigma} \cdot [\boldsymbol{\Pi}^{\mathrm{d}} + g_1(\alpha)\boldsymbol{\Pi}^{\mathrm{c}}]\boldsymbol{\sigma}} - g_2(\alpha)\sigma_{\mathrm{y}} = 0.} \tag{3.26}$$

Constitutive functions $g_i(\alpha)$ in (3.26) are specified later. It is, however, required that, for completely flattened asperities (i.e. for $\alpha = 1$), the yield condition (3.26) of the surface layer reduces to the yield condition $(3.13)_1$ of the bulk material. This implies that $g_1(1) = 0$ and $g_2(1) = 1$.

Remark 3.1. The associated flow rule (3.24) with $\boldsymbol{\Pi}^{\mathrm{l}}$ defined by (3.25) generates non-zero volumetric strain. Thus, in the present modelling, the surface layer can be treated as a fictitious porous body. The porosity is related to surface roughness and decreases in the deformation process under the action of contact tractions. The surface layer exhibits thus configurational hardening due to decreasing porosity until the strength of the bulk material is recovered for completely flattened asperities.

3.3.4 Evolution Law for the Real Contact Area Fraction

Following the approach presented in Sect. 3.2.3, the strain rate and the stress in the surface layer can be expressed in terms of the macroscopic quantities, specifically in terms of $\Sigma_A = \sigma_A$ and $\dot{E}_P = \dot{\varepsilon}_P$. The yield condition (3.26) can be transformed to the mixed form analogous to that given by (3.20), and, for the particular form (3.25) of $\boldsymbol{\Pi}^l$, the yield condition can be written as

$$F_v^l(P_N, P_T, E_v, \alpha) = \sqrt{\frac{g_1}{3g_2^2}\left[1 + g_1(1 - P_T^2)E_v^2\right]P_N^2 + \frac{1+g_1}{g_2^2}P_T^2} - 1 = 0,$$

(3.27)

where P_N and P_T are the dimensionless contact stresses, and E_v is the dimensionless macroscopic (bulk) strain rate, i.e. the macroscopic equivalent strain rate $\dot{E}$ normalized by the volumetric strain rate $\dot{\varepsilon}_v$, namely

$$E_v = \frac{\dot{E}}{\dot{\varepsilon}_v}, \qquad \dot{E} = \sqrt{\frac{2}{3}\,\dot{\mathbf{E}} \cdot \dot{\mathbf{E}}}, \qquad \dot{\varepsilon}_v = -\mathrm{tr}\,\dot{\varepsilon} = \frac{g_1 P_N}{\sqrt{3}g_2}\,\gamma^l. \qquad (3.28)$$

Note that the macroscopic equivalent strain rate $\dot{E}$ enters the yield condition (3.27) through equation (3.19) relating $\dot{E}$ and its interior part $\dot{E}_P$. Using (3.28), it can be checked that for $g_1 = 0$ and $g_2 = 1$, i.e. for $\alpha = 1$, the yield condition (3.27) becomes an identity, since for $\alpha = 1$ we have $\gamma^l = \dot{E}$.

In (3.27), the components of Σ_A are replaced by the components of the contact traction vector $\mathbf{T} = \Sigma\mathbf{n}$ ($T_N = \Sigma_{33}$, $T_{T1} = \Sigma_{13}$, $T_{T2} = \Sigma_{23}$). The normal contact pressure and the friction stress are then normalized using the yield stress in shear, $k = \sigma_y/\sqrt{3}$, so that

$$P_N = -\frac{T_N}{k}, \qquad P_T = \frac{T_T}{k}, \qquad T_T = \sqrt{T_{T1}^2 + T_{T2}^2}. \qquad (3.29)$$

In order to close the model, the real contact area fraction α must be related to the deformation of the surface layer. Based on simple geometric considerations the following relation has been proposed by Stupkiewicz and Mróz [138],

$$\dot{\alpha} = \frac{\dot{\varepsilon}_v}{\varepsilon_v^{max}}, \qquad \varepsilon_v^{max} = \log\frac{\eta + \tan\theta}{\eta + \nu\tan\theta} \approx \frac{(1-\nu)\theta}{\eta}, \qquad (3.30)$$

where ε_v^{max} is the maximum volumetric strain in the surface layer. Parameters θ and ν depend on the initial roughness: θ is the average asperity slope; and $1-\nu$ is the "porosity" of the layer contained between the highest asperity summits and the deepest asperity valleys, for instance, $\nu = \frac{1}{2}$ in the case of two-dimensional wedge-like asperities. The phenomenological parameter η denotes the thickness of the surface layer relative to the characteristic asperity spacing. Alternatively, ε_v^{max} can directly be used as a phenomenological parameter of the model without specifying parameters θ, ν, and η.

The dimensionless macroscopic strain rate E_v can now be determined from the yield condition (3.27), viz.

$$E_v = \hat{E}_v(P_N, P_T, \alpha) = \sqrt{\frac{3g_2^2 - g_1 P_N^2 - 3(1 + g_1)P_T^2}{g_1^2 P_N^2(1 - P_T^2)}}. \qquad (3.31)$$

Finally, using $(3.28)_1$, $(3.30)_1$, and (3.31), the evolution law for the real contact area fraction is obtained in the form

$$\dot{\alpha} = \frac{\dot{E}}{\varepsilon_v^{\max} \hat{E}_v(P_N, P_T, \alpha)}. \qquad (3.32)$$

According to this evolution law, $\dot{\alpha}$, the rate of the real contact area fraction, is proportional to the macroscopic equivalent plastic strain rate $\dot{E}$, and the proportionality factor depends on the dimensionless contact stresses P_N and P_T, and on the real contact area fraction α.

3.3.5 Specification of Constitutive Functions $g_i(\alpha)$

The constitutive functions $g_i(\alpha)$ must satisfy some physical constraints. Firstly, as already discussed, $g_1(1) = 0$ and $g_2(1) = 1$. Secondly, in the case of no macroscopic deformation ($E_v = 0$), we require that, at $\alpha = 0$, the effective dimensionless hardness H, defined as

$$H = \frac{P_N}{\alpha}, \qquad (3.33)$$

is equal to $H = 2 + \pi$, which is a classical result of the rigid-plastic indentation problem, cf. Hill [44], Szczepiński [148]. The following functions g_1 and g_2,

$$g_1(\alpha) = 1 - \alpha, \qquad g_2(\alpha) = \alpha^2 + \frac{2 + \pi}{\sqrt{3}}\alpha(1 - \alpha), \qquad (3.34)$$

satisfy the above conditions, and have been found to provide satisfactory prediction at zero macroscopic strain, as illustrated below.

Consider thus flattening of asperities in frictionless conditions, $P_T = 0$, and in the absence of macroscopic deformation, $E_v = 0$. The following relation between P_N and α is obtained from the asperity flattening condition (3.27),

$$P_N = \sqrt{3}\frac{g_2(\alpha)}{\sqrt{g_1(\alpha)}}. \qquad (3.35)$$

Figure 3.3 presents the real contact area fraction α as a function of the dimensionless contact pressure P_N, as predicted by (3.34) and (3.35). This relation is compared to the respective prediction of the micromechanical model of Wanheim et al. [156], and a good agreement of the two models is clearly seen in Fig. 3.3. It is shown in Stupkiewicz and Mróz [138] that the present model reasonably agrees with the model of Wanheim et al. [156] also in the case of non-zero friction, $P_T \neq 0$.

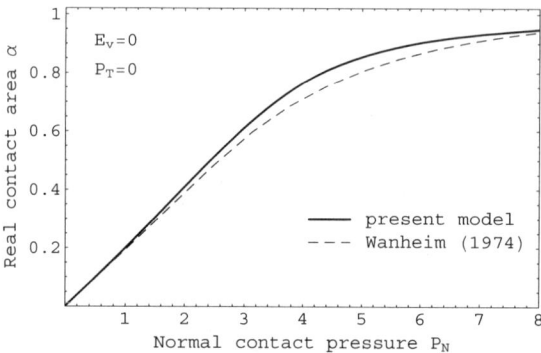

Fig. 3.3. Real contact area fraction α as a function of dimensionless normal contact pressure P_N in the case of zero macroscopic strain, $E_v = 0$ (originally published in [138], copyright Elsevier).

3.3.6 Model Predictions

The evolution law (3.32) for the real contact area fraction α, resulting from the present model, has been successfully verified against the existing micromechanical models and available experimental data, see Stupkiewicz and Mróz [138]. Selected results are provided below.

Wilson and Sheu [162] analyzed a two-dimensional periodic indentation problem of a macroscopically deforming rigid-plastic half-space indented by a periodic array of rigid indenters. The solution, obtained using the upper bound method, has been approximated with an analytical function relating the effective hardness H to the real contact area fraction α and dimensionless macroscopic strain rate E_f. The effective hardness H, cf. (3.33), has an interpretation of the average contact pressure at real asperity contacts. The dimensionless macroscopic strain rate $E_f = \dot{E}l/v_f$ is the macroscopic equivalent plastic strain rate $\dot{E}$ normalized using the flattening velocity v_f and the characteristic asperity spacing l. Assuming that the asperities have the form of wedges with constant asperity slope θ, it follows that

$$E_f = \frac{1-\nu}{\eta}\,E_v = \frac{1}{2\eta}\,E_v, \tag{3.36}$$

where, for wedge-like asperities, we have $\nu = \frac{1}{2}$, and the flattening velocity is given by $v_f = \dot{\alpha} l \tan\theta$.

In Figure 3.4, predictions of the present model, corresponding to $\eta = 1$, are compared to predictions of the micromechanical model of Wilson and Sheu [162]. The agreement is considered satisfactory, particularly, in view of the idealized asperity layout assumed by Wilson and Sheu [162], and in view of other assumptions of that model.

The present model predicts that the effective hardness decreases with increasing dimensionless macroscopic strain rate E_f, cf. Fig. 3.4. This is also

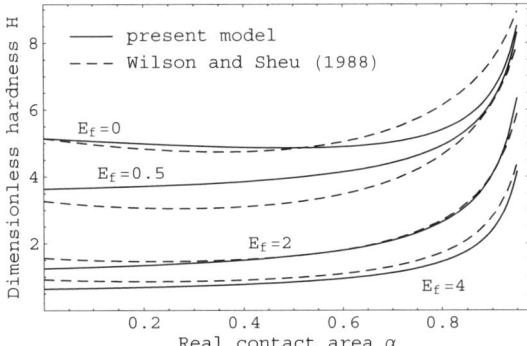

Fig. 3.4. Effective hardness H as a function of the fraction α of real contact area (originally published in [138], copyright Elsevier).

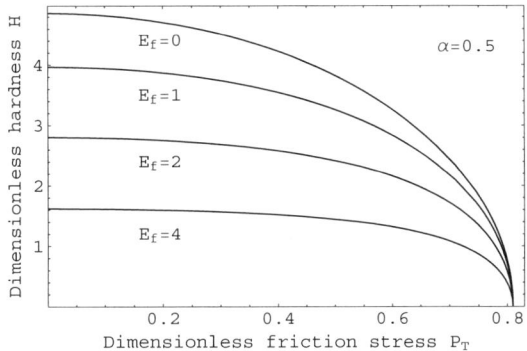

Fig. 3.5. Effective hardness H as a function of the friction stress P_T for $\alpha = 0.5$.

seen in Fig. 3.5, where the hardness H is shown as a function of macroscopic friction stress P_T. According to the present model, friction induces additional decease of the effective hardness.

Predictions of the present model have also been compared to the experimental data of Sutcliffe [147] and Wilson and Sheu [162]. In the asperity crushing experiment of Sutcliffe [147], copper bars with ridges machined across opposite faces were compressed on the ridged surfaces by two flat and smooth dies and, subsequently, stretched in the perpendicular direction, cf. Fig. 3.6. Two series of experiments were conducted with ridges perpendicular to the straining direction (transverse roughness, as illustrated in Fig. 3.6) and with ridges parallel to the straining direction (longitudinal roughness). A constant macroscopic normal pressure was prescribed, and at each load increment the real contact area fraction was measured. Geometry of the specimen was appropriate to adopt the plane strain assumption. The case of longitudinal roughness was also studied experimentally by Wilson and Sheu [162] who measured the real area of contact as a function of the macroscopic strain in rolling.

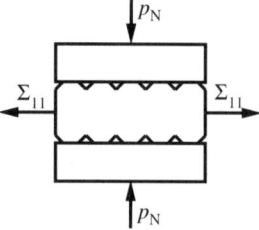

Fig. 3.6. Asperity crushing experiment of Sutcliffe [147].

Once the history of contact tractions (here, P_N = const and P_T = 0) is prescribed, the evolution equation for α, cf. (3.32), can be integrated to yield the real contact area fraction α as a function of macroscopic deformation; in the present case, as a function of macroscopic strain E_{11}. The model involves one adjustable parameter, namely the relative thickness of the surface layer η, while ν and θ are known. The values of parameter η, which provide best fit of each set of experimental data, are given in Table 3.1. The values corresponding to transverse roughness are 2–3 times higher than those corresponding to longitudinal roughness. This is in a qualitative agreement with the finite element solutions of Korzekwa et al. [69] who predicted minimum effective hardness for transverse roughness and significantly higher effective hardness for longitudinal roughness. The values of η providing the best fit of experimental data for highly anisotropic roughness formed by two-dimensional ridges are between 1.2 and 1.88 for bulk straining direction perpendicular to the ridges (transverse roughness) and between 0.46 and 0.68 in the case of longitudinal roughness. The value of parameter η in the case of isotropic asperity layout is thus expected to be close to unity.

Table 3.1. Thickness parameter η providing best fit of experimental data.

	P_N	η	ε_v^{max}
Transverse roughness, Sutcliffe [147]	0.8	1.50	0.12
	1.32	1.88	0.09
	2.0	1.20	0.15
Longitudinal roughness, Sutcliffe [147]	1.2	0.46	0.38
	2.0	0.48	0.36
Longitudinal roughness, Wilson & Sheu [162]	2.0	0.68	0.13

In Figure 3.7, the predicted growth of the real contact area with increasing macroscopic strain E_{11} is compared to the transverse roughness results of Sutcliffe [147]. The model predictions in Fig. 3.7 correspond to only one value of parameter η, the value which provides the best fit for P_N = 2, namely $\eta = 1.20$.

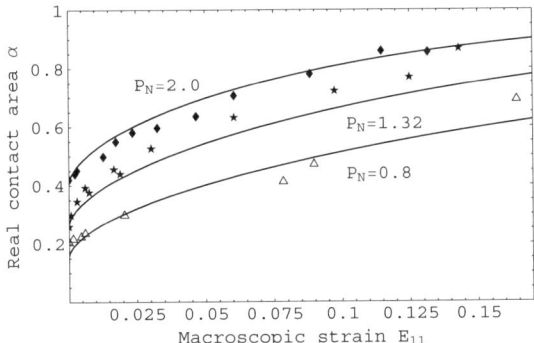

Fig. 3.7. Real contact area fraction α as a function of macroscopic strain E_{11}. Transverse roughness results of Sutcliffe [147] and model predictions for $\eta = 1.20$ (originally published in [138], copyright Elsevier).

Weakening due to asperity flattening and due to related inhomogeneity of plastic deformation in the surface layer plays a central role in the present modelling. In practice, the typical asperity dimensions and the thickness of the surface layer are small compared to the dimensions of the workpiece. Thus weakening of the surface layer has a negligible effect on the overall deformation of the workpiece, both in bulk and sheet forming operations. However, in the experiments of Sutcliffe [147], the initial thickness of the bar, $2h_0 = 10\,[\mathrm{mm}]$, was only four times greater than the initial ridge spacing, $2l_0 = 2.6\,[\mathrm{mm}]$, so that the weakening effect of deformation inhomogeneity was observed and measured.

Assuming that the bar is in tension, $\dot{E}_{11} > 0$, the plane strain yield condition of the bulk material is given by $\Sigma_{11} - \Sigma_{33} = 2k$, which corresponds to the frictionless case, $\Sigma_{13} = T_{T1} = 0$. Similarly, the stress in the surface layer can be obtained from the yield condition (3.23) using the flow rule (3.24) corresponding to the plane strain conditions, $\dot{\varepsilon}_{22} = 0$, so that

$$\Sigma_{11}/2k = 1 - P_N/2, \quad \sigma_{11}/2k = \sqrt{[g_2(\alpha)]^2 - \frac{1}{3}g_1(\alpha)P_N^2} - P_N/2. \quad (3.37)$$

The average stress in the bar, $\Sigma_{11}^{\mathrm{av}}$, can now be estimated from the following expression,

$$\Sigma_{11}^{\mathrm{av}} = \frac{h^b \Sigma_{11} + h^l \sigma_{11}}{h^b + h^l}, \quad (3.38)$$

where $h^l = (\eta + \tan\theta)l_0$ is the thickness of the surface layer, and $h^b = h_0 - h^l$ is the thickness of the remaining bulk material. For simplicity, the variation of h^l and h^b during deformation of the bar has been neglected in (3.38), as this variation has a minor effect on $\Sigma_{11}^{\mathrm{av}}$.

Figure 3.8 presents the real contact area fraction α as a function of the sum of applied stresses, $P_N/2 + \Sigma_{11}/2k$. The prediction of the present model

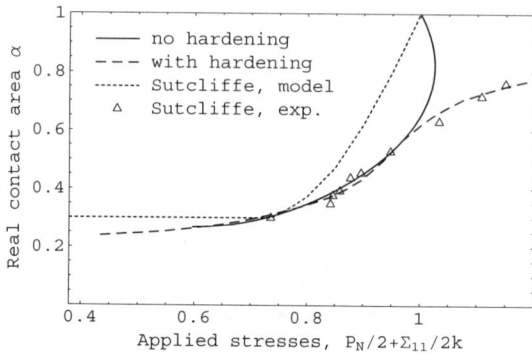

Fig. 3.8. Relation between the real contact area fraction α and the sum of applied stresses needed to flatten transverse asperities under constant normal pressure $P_N = 1.32$.

corresponding to $\eta = 1.20$ (solid line in Fig. 3.8), as specified by (3.37) and (3.38), is compared to the experimental data ($P_N = 1.32$) and to the slip-line solution of Sutcliffe [147].

Both the present model and the Sutcliffe's model predict that for $\alpha = 1$, i.e. for completely flattened asperities, the sum of normalized applied stresses is equal to unity, which is a correct result corresponding to a homogeneous deformation in plane strain conditions. Note, however, that the experimentally measured stresses, normalized here by the *initial* yield stress k, are significantly higher than unity. This has been explained by Sutcliffe [147] to result from strain hardening. Indeed, a simple extension of the present model to account for the strain hardening effects provides an excellent agreement with experimental data, cf. Krasniuk and Stupkiewicz [72]. The corresponding diagram is also included in Fig. 3.8 (dashed line). Note that the hardening curve of the bulk material, which has been measured and reported by Sutcliffe [147], is the only additional input required by the extended model of Krasniuk and Stupkiewicz [72].

3.3.7 Friction Model

As an application of the evolution law (3.32), a friction model is briefly outlined below. The model, proposed by Stupkiewicz and Mróz [136, 137], is based on the assumption that the tool asperities are much smaller than the workpiece asperities, so that the combined asperity interaction mode, cf. Mróz and Stupkiewicz [84], Wilson [160], can be adopted, cf. Fig. 3.9.

The average dimensionless local contact stresses at workpiece asperity contacts are given by

$$P_N^a = \frac{P_N}{\alpha_w}, \qquad P_T^a = \frac{P_T}{\alpha_w}, \tag{3.39}$$

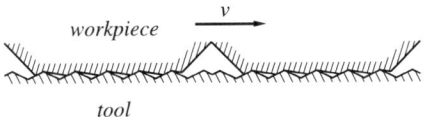

Fig. 3.9. Combined asperity interaction mode: workpiece asperities are flattened according to the RW-ST interaction mode, and much smaller tool asperities plough through the plateaus of the flattened workpiece asperities.

where α_w is the measure of flattening of workpiece asperities. Asperity flattening is assumed to proceed according to the RW-ST interaction mode, cf. Fig. 3.1(a), and the evolution law for α_w is thus specified by (3.32) with α replaced by α_w. Note that, in the assumed two-scale asperity interaction mode, the real contact area fraction is given by $\alpha = \alpha_w \alpha_t$, where α_t is the local real contact area fraction at the workpiece asperity plateaus, according to the SW-RT mode in Fig. 3.1(b), see also Mróz and Stupkiewicz [84].

Frictional interaction at workpiece asperity contacts is assumed to be governed by adhesion and ploughing of tool asperities through the plateaus of flattened workpiece asperities. The corresponding local friction law is assumed in the form of a simple nonlinear law, cf. Stupkiewicz [130],

$$f^{a}(P^{a}_{N}, P^{a}_{T}) = |P^{a}_{T}| - m \tanh\left(\frac{\mu P^{a}_{N}}{m}\right) \leq 0, \tag{3.40}$$

where μ is the friction coefficient at low contact pressures, and m is the so-called friction factor, so that $P^{a}_{T} = m$ at high contact pressures. Combining (3.40) and (3.39), the following limit friction condition is obtained

$$f(P_{N}, P_{T}, \alpha_w) = |P_{T}| - \alpha_w m \tanh\left(\frac{\mu P_{N}}{\alpha_w m}\right) \leq 0. \tag{3.41}$$

This condition involves a state variable α_w with evolution law specified by (3.32). Note that α_w is not a unique function of the current contact state, but it depends on the history. Surface roughening due to plastic deformations, cf. Wilson and Lee [161], is not accounted for by the evolution law (3.32), thus α_w increases monotonically. With increasing α_w, the limit friction surface $f = 0$ grows in a self-similar manner, as schematically illustrated in Fig. 3.10.

As the applicability of the present model for finite element simulations is concerned, it should be noted that the macroscopic equivalent plastic strain rate $\dot{E}$, occurring in the evolution law (3.32) for α_w, is a non-standard contact variable that cannot be handled by the usual finite element formulations of contact. A possible approach to treat this problem has been proposed by Stupkiewicz and Mróz [136]. The approach relies on the relation (3.19) between the macroscopic equivalent plastic strain rate $\dot{E}$ and its interior contribution $\dot{E}_P$. The latter one can be, in fact, determined from surface data. For instance, in plane strain conditions ($\phi = 0$), the increment of the (logarithmic) in-plane

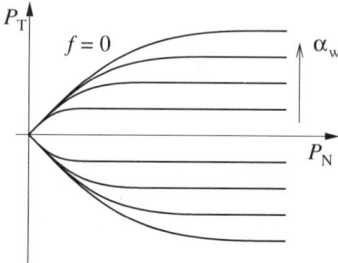

Fig. 3.10. Evolution of the limit friction surface.

strain component ΔE_{11} and the increment of total equivalent plastic strain ΔE can be approximated by

$$\Delta E_{11} = \log(A^{n+1}/A^n), \qquad \Delta E = \frac{2|\Delta E_{11}|}{\sqrt{3(1 - P_T^2)}}, \qquad (3.42)$$

where A^{n+1} and A^n denote the equivalent area of a contact integration point at, respectively, the current time increment and at the previous one. The above approximation can effectively be handled using the *extended node-to-segment contact element* developed by Stupkiewicz [130]. Results of a representative finite element simulation, employing the friction model (3.41) with the evolution law (3.32), can be found in Stupkiewicz and Mróz [136].

3.4 Conclusions

An approach to describe the effect of macroscopic (bulk) plastic deformations on asperity flattening and on evolution of real contact area in metal forming processes has been presented in this chapter. The phenomenological framework, presented first by Stupkiewicz and Mróz [138], is based on the assumption that the surface asperities and the underlying thin layer of inhomogeneous deformation can be represented by an equivalent homogeneous surface layer. A micromechanical approach to treat inhomogeneous deformations within contact layer is developed in Chaps. 4, 5, and 6.

As the first step of the present modelling, rigid-plastic constitutive equations are postulated for the surface layer. The yield condition of the surface layer, is then expressed in the mixed form, i.e. in terms of macroscopic quantities only. This condition, together with the heuristic postulate (3.30), provides a closed-form relationship between the rate of the real contact area fraction, the contact state variables (the contact stresses and the real contact area fraction), and the macroscopic equivalent plastic strain rate. The effective hardness predicted by the model increases with increasing real contact area fraction and decreases with increasing dimensionless macroscopic equivalent plastic strain rate. Despite many simplifying assumptions, the model provides a good agreement with existing micromechanical models and with available experimental data.

The model predicts that, under constant normal pressure, the real contact area grows asymptotically to unity with increasing bulk strain, cf. Fig. 3.7. This is consistent with the predictions of the model of Wilson and Sheu [162]. However, qualitatively different behaviour is predicted by Sutcliffe [147] and also by Kimura and Childs [61]. The model of Sutcliffe [147] predicts that $\alpha = 1$ is attained at a finite value of bulk strain. On the other hand, persistence of longitudinal asperities is predicted by Kimura and Childs [61]. Their results suggest that there exists a limit real contact area fraction (which depends on the contact conditions, e.g. friction) between 0.75 and 0.95 according to their theory (based on the upper bound method) and between 0.6 and 0.8 according to the experiment. Each of the two effects might be included in the present model by refining the evolution law for the real contact area. However, as the effects are opposite, additional experimental evidence would first be required.

The model involves one adjustable parameter, namely the relative thickness η of the surface layer. Alternatively, if roughness characteristics (average asperity slope θ and "porosity" parameter ν) are not known, then the maximum volumetric strain $\varepsilon_v^{\max}$ can be adopted as a phenomenological parameter. The expected value of parameter η in the case of isotropic asperity layout is close to unity, as the values providing the best fit of experimental data for highly anisotropic roughness formed by two-dimensional ridges are between 1.2 and 1.88 for transverse roughness and between 0.46 and 0.68 for longitudinal roughness.

The model can easily be extended to account for the effects of strain hardening in the bulk material and in the surface layer by adopting respective yield conditions with isotropic-type hardening, cf. Krasniuk and Stupkiewicz [72]. The resulting structure of the model is similar to that presented in Sect. 3.3. While the predicted evolution of real contact area is not much affected by strain hardening, the weakening effect can be modelled more accurately, cf. Fig. 3.8.

4

Boundary Layers Induced by Micro-Inhomogeneous Boundary Conditions

Abstract: Boundary layers induced by micro-inhomogeneous boundary conditions are studied. The notion of macro-scale and micro-scale is introduced and the method of asymptotic expansions is used to derive the equations of the corresponding macroscopic and microscopic boundary value problems for an elastic body in two dimensions. Three specific cases of micro-inhomogeneous boundary conditions are considered in detail: prescribed surface traction, prescribed displacement, and frictional contact of a rough body with a rigid and smooth obstacle.

4.1 Motivation

In the case of contact of rough bodies, the characteristic dimension of roughness is typically much smaller than that of the contacting bodies, see Fig. 4.1. Thus two points of view can be adopted. At the *micro-scale*, the stress transfer is concentrated at small spots, so-called real contacts, and the distribution of contact traction is highly inhomogeneous. These inhomogeneities govern the interaction and deformation of surface asperities. Furthermore, a thin subsurface layer of inhomogeneous deformation is induced. At the *macro-scale*, it is the slowly-varying average (macroscopic) contact traction that determines the overall deformation of the contacting bodies.

While frictional interactions are governed by the *local* phenomena at the micro-scale, the friction laws are typically formulated in terms of the normal and tangential components of the *macroscopic* contact traction vector $\mathbf{t} = \boldsymbol{\sigma}\mathbf{n}$, where $\boldsymbol{\sigma}$ is the macroscopic stress tensor at the contact surface, and $\mathbf{n}$ is the unit outward normal. Consequently, only the *exterior* part of the stress tensor $\boldsymbol{\sigma}$, cf. Sect. 2.2, is involved in the description, and the complete stress and strain state in the vicinity of the contact surface is typically not accounted for. However, in some situations, the *interior* (in-plane) parts of stress or strain significantly affect friction and other contact phenomena, particularly, if the

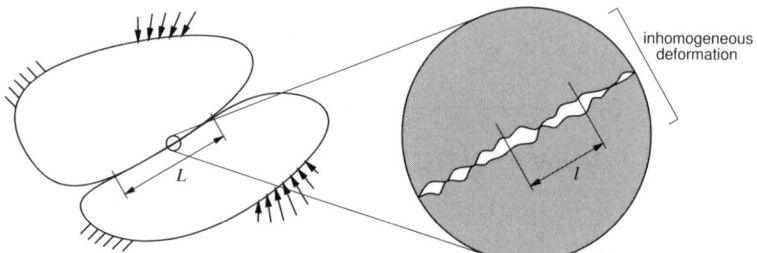

Fig. 4.1. Contact of rough bodies: macro- and micro-scale.

deformation in the subsurface layer is inhomogeneous. This is, for example, the case of metal forming processes where the surface asperities are flattened more easily in the presence of macroscopic plastic deformation. This leads to high real contact area fractions, even at moderate contact pressures; refer to Sect. 3.3 for a more detailed discussion. Depending on the lubrication regime, this can result in an increased adhesive friction component or affect lubrication conditions. A closely related effect is also observed in hardness indentation testing, where the in-plane stresses affect the force-penetration response, cf. Giannakopoulos [33].

It seems that the effects associated with deformation inhomogeneities within subsurface contact layers and the interaction of these inhomogeneities with the macroscopic stresses and strains have not attracted sufficient attention in the literature yet. The aim of this chapter is thus to develop a micromechanical framework that would allow a consistent analysis of these effects and, in a broader perspective, would help to develop improved constitutive laws of contact phenomena.

The importance of surface roughness in the mechanics of contact interactions is, of course, very well recognized. The area of real contact of rough bodies, which is directly related to interaction of surface asperities, usually constitutes only a small fraction of the nominal contact area, and the related effects at the micro-scale govern the contact phenomena (friction, wear, contact compliance, heat transfer, lubrication, etc.) observed at the macro-scale.

For instance, local deformations of deforming asperities and the related contact compliance may, in some situations, affect the overall deformations of rough bodies in contact. Contact compliance of a rough surface, or of a pair of rough surfaces, can be obtained by determining the response of a single (representative) asperity, and by subsequent averaging over statistical distribution of asperity heights, radii, etc., cf. Greenwood and Williamson [34], Greenwood and Wu [35], Whitehouse and Archard [158], Kucharski et al. [74]. As a result, Winkler-type contact laws are obtained, which relate the contact pressure and the relative approach of the nominal surfaces. Contact laws of this type can be readily used in finite element computations, e.g. Buczkowski and Kleiber [19],

Oden and Martins [88], Wriggers et al. [166]. Another approach to include the effect of roughness into macro modelling of contact interactions has been proposed by Pauk and Woźniak [98]. In that approach, so-called micro-shape and decay functions, which are assumed *a priori*, are used to model inhomogeneous displacements at the micro-scale, see also Woźniak [164].

In this chapter, the boundary layer approach is used to derive equations governing the deformation in the subsurface layer in the limit of $\varepsilon = l/L \to 0$, where l is the characteristic dimension of the inhomogeneities, e.g. the characteristic dimension of roughness, and L is the characteristic dimension of the macroscopic contact zone, cf. Fig. 4.1. The resulting boundary value problem of the boundary layer is formulated for a half-space subjected to some periodic loading, e.g. surface traction, contact interaction, etc. In fact, the available micromechanical models of asperity interaction are often based on the analysis of a half-space with a periodic arrangement of asperities, e.g. Avitzur and Nakamura [5], Sutcliffe [147], Wanheim et al. [156], Wilson and Sheu [162], and many others. In all those cases, the simplified problem of a rough half-space is introduced in an intuitive way. The boundary layer analysis presented below allows derivation of the governing equations in a more rigorous way, and, importantly, provides the framework for the analysis of effects of macroscopic stresses and strains, which is carried out in Chaps. 5 and 6.

The boundary layer analysis is a classical approach within the field of micromechanics and homogenization of heterogeneous materials, see, for instance, Sanchez-Palencia [125], Pruchnicki [108], Luciano and Willis [81]. In composite materials, the boundary layers appear because the micro-periodic solutions, predicted by the homogenization theory, are not valid in the vicinity of the boundaries. This is because the micro-periodic solutions satisfy the boundary conditions only in an average sense, and, secondly, the assumption of periodicity in the direction normal to the boundary is no longer valid. These effects are accounted for by considering additional correction terms in the mechanical fields in the vicinity of boundaries and edges.

A different application of the boundary layer analysis is presented in this chapter. Boundary layers are considered, which are induced in *homogeneous bodies* subjected to *micro-inhomogeneous boundary conditions*, as, for instance, in the case of contact of rough bodies. The formalism of boundary layer analysis, adopted below, is based on that of Sanchez-Palencia [125].

The content of the present chapter is the following. Boundary layers associated with a micro-inhomogeneous prescribed surface traction are introduced in Sect. 4.2. Next, in Sect. 4.3, the case of micro-inhomogeneous prescribed displacements is briefly discussed. Finally, contact and friction conditions are analyzed in Sects. 4.4 and 4.5 for the case of a rough body in contact with a smooth rigid obstacle.

4.2 Micro-Inhomogeneous Surface Traction

4.2.1 Problem Statement

Consider a homogeneous body occupying domain Ω, and assume that the surface traction $\mathbf{t}^b$ prescribed on the boundary Γ_t is *micro-inhomogeneous*. By micro-inhomogeneity of $\mathbf{t}^b$ we understand that it consists of a slowly varying average field $\bar{\mathbf{t}}^b$ and its micro-periodic fluctuation $\tilde{\mathbf{t}}^b$. The wave-length of the fluctuation field l is assumed small compared to the length L of the boundary Γ_t and small compared to the overall dimensions of the body. In order to keep the exposition and the notation as simple as possible, the discussion in this chapter is restricted to a two-dimensional elasticity problem – in order to fix the attention let us assume the plane strain conditions.

The *background boundary value problem* is thus to find the displacement field $\mathbf{u}(\mathbf{x})$ satisfying the following set of equations

$$\begin{cases} \operatorname{div}\boldsymbol{\sigma} = \mathbf{0}, \ \mathbf{x} \in \Omega \\ \boldsymbol{\sigma} = \mathbf{L}\mathbf{e}(\mathbf{u}), \mathbf{x} \in \Omega \\ \boldsymbol{\sigma}\mathbf{n} = \mathbf{t}^b, \quad \mathbf{x} \in \Gamma_t \\ \mathbf{u} = \mathbf{u}^b, \quad \mathbf{x} \in \Gamma_u \end{cases} \tag{4.1}$$

where

$$\mathbf{e}(\mathbf{u}) = \frac{1}{2}[\operatorname{grad}\mathbf{u} + (\operatorname{grad}\mathbf{u})^T] \tag{4.2}$$

is the infinitesimal strain, $\mathbf{n}$ the unit outward normal, $\mathbf{u}^b$ the prescribed displacement, and the micro-periodic surface traction $\mathbf{t}^b$ is given by

$$\mathbf{t}^b(\xi) = \bar{\mathbf{t}}^b(\xi) + \tilde{\mathbf{t}}^b(\xi) = \bar{\mathbf{t}}^b(\xi) + \mathbf{t}_a^b(\xi)\,\tilde{p}(\xi), \tag{4.3}$$

where $\xi \in (0, L)$ is the macroscopic arc-length parameter that parameterizes the boundary Γ_t. The local average $\bar{\mathbf{t}}^b(\xi)$ and the local amplitude $\mathbf{t}_a^b(\xi)$ of the traction $\mathbf{t}^b$ are slowly varying functions while $\tilde{p}(\xi)$ is a periodic, purely oscillating function,

$$\tilde{p}(\xi) = \tilde{p}(\xi + l), \qquad \int_\xi^{\xi+l} \tilde{p}(\xi')\mathrm{d}\xi' = 0. \tag{4.4}$$

As the material is assumed to be homogeneous, the elastic stiffness tensor $\mathbf{L}$ does not depend on the position within Ω.

Direct solution of the problem defined above, e.g. by the finite element method, may be difficult or even impossible because of the two scales involved, so that discretization accounting for the fluctuations of surface traction would be prohibitively fine. As an alternative, the boundary layer analysis can be carried out. The idea of this approach is briefly outlined below.

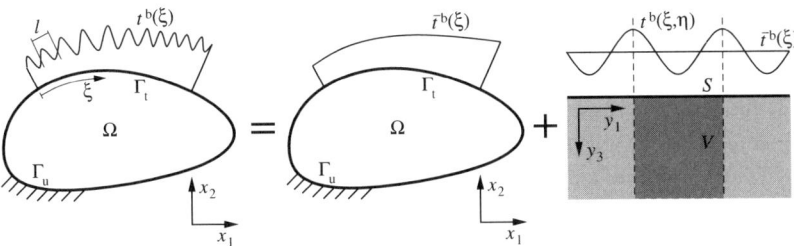

Fig. 4.2. Boundary layers induced by micro-inhomogeneous boundary conditions: the idea of the proposed approach.

By the Saint-Venant's principle, inhomogeneity of deformation induced by the micro-inhomogeneous boundary condition is confined to a thin subsurface layer along Γ_t, while the thickness of this layer is of the order of l, the fluctuation period of $\mathbf{t}^b$.

The *macroscopic problem*, i.e. a boundary value problem corresponding to the average, micro-homogeneous surface traction $\bar{\mathbf{t}}^b$, can thus be posed. However, the macroscopic stresses, following from the solution of the macroscopic problem, satisfy the micro-inhomogeneous boundary condition $(4.1)_3$ of the background problem in the average only. A correction of the solution in the subsurface layer is thus necessary. Within the boundary layer approach, this correction follows from the *microscopic problem*, which is obtained by scaling the coordinates and then considering the form that the equations take as $\varepsilon = l/L$ tends to zero. The background problem with micro-inhomogeneous surface traction is thus replaced by the macroscopic problem with micro-homogeneous surface traction accompanied by the microscopic problem at each point $\mathbf{x} \in \Gamma_t$, cf. Fig. 4.2. Importantly, the microscopic problem is formulated for a simplified geometry, namely for a periodically loaded half-space. Details of the above scheme are provided in subsequent sections.

The present approach is largely based on that of Sanchez-Palencia [125], who analyzed boundary layers and edge effects in composites. Also, the mathematical background of the asymptotic methods, which are extensively used in the homogenization theory and which are a basic tool of the present analysis, can be found in [125].

4.2.2 Macroscopic Problem

Let us first formulate the *macroscopic problem* with the unknown macroscopic displacement field $\mathbf{u}^0(\mathbf{x})$, viz.

$$\begin{cases} \operatorname{div} \boldsymbol{\sigma}^h = \mathbf{0}, & \mathbf{x} \in \Omega \\ \boldsymbol{\sigma}^h = \mathbf{Le}(\mathbf{u}^0), & \mathbf{x} \in \Omega \\ \boldsymbol{\sigma}^h \mathbf{n} = \bar{\mathbf{t}}^b, & \mathbf{x} \in \Gamma_t \\ \mathbf{u}^0 = \mathbf{u}^b, & \mathbf{x} \in \Gamma_u \end{cases} \qquad (4.5)$$

which corresponds to the *micro-homogeneous* surface traction $\bar{\mathbf{t}}^b$ on Γ_t, and $\boldsymbol{\sigma}^h(\mathbf{x})$ denotes the macroscopic stress.

Considering the background problem (4.1), solution $\mathbf{u}^0(\mathbf{x})$ of the macroscopic problem (4.5) is not a solution of the background problem as the boundary condition (4.1)$_3$ is not satisfied by $\mathbf{u}^0(\mathbf{x})$, though it is satisfied in the average sense. Thus a *boundary layer* must appear along Γ_t with extra terms in the displacement field. The solution $\mathbf{u}^0(\mathbf{x})$ of the macroscopic problem (4.5) is thus valid everywhere within Ω, except in the vicinity of boundary Γ_t.

4.2.3 Two-Scale Description and Asymptotic Expansions

In order to study the micro-inhomogeneous fields within the boundary layer, two spatial variables are introduced in the usual spirit of the homogenization theory: the macro variable $\mathbf{x}$ and the micro variable $\mathbf{y}$, the latter is scaled according to

$$\mathbf{y} = \mathbf{x}/\varepsilon, \tag{4.6}$$

where $\varepsilon = l/L \ll 1$ is a small parameter.

Next, the asymptotic expansion of the displacement field is introduced in the vicinity of the boundary Γ_t, the corresponding part of Ω is denoted by Γ_t^+, so that

$$\mathbf{u}^\varepsilon(\mathbf{x}) = \mathbf{u}^0(\mathbf{x}) + \varepsilon\mathbf{u}^1(\mathbf{x}',\mathbf{y}) + O(\varepsilon^2), \quad \mathbf{x} \in \Gamma_t^+, \; \mathbf{x}' \in \Gamma_t, \; \mathbf{y} \in V, \tag{4.7}$$

where $\mathbf{u}^1(\mathbf{x}',\mathbf{y})$ is the first-order boundary layer correction term, $\mathbf{x}'$ is the orthogonal projection[1] of $\mathbf{x}$ onto Γ_t, and $O(\varepsilon^n)$ is the classical order symbol, so that $f = O(\varepsilon^n)$ means that $|f| < A\varepsilon^n$ for some constant A. The micro variable $\mathbf{y}$ runs in the strip V, cf. Fig. 4.3,

$$V = \{(y_1, y_3) \colon y_1 \in (0, L); \; y_3 \in (0, +\infty)\}, \tag{4.8}$$

and it is represented in the intrinsic coordinate system with its y_3-axis normal to the boundary and directed inwards.

The displacement correction term $\mathbf{u}^1$ is *V-periodic*, i.e. periodic with respect to $y_1 = \eta$ but not with respect to y_3,

$$\boxed{\mathbf{u}^1(\mathbf{x}',\mathbf{y}^-) = \mathbf{u}^1(\mathbf{x}',\mathbf{y}^+),} \tag{4.9}$$

where $\mathbf{y}^- = (0, y_3)$ and $\mathbf{y}^+ = (L, y_3)$, cf. Fig. 4.3, and the period in the micro variable y_1 is $L = l/\varepsilon$. Accordingly, the expansion (4.7) is *almost periodic* in the micro variable y_1, cf. Sanchez-Palencia [125]. Furthermore, the gradient of $\mathbf{u}^1$ with respect to the micro variable $\mathbf{y}$ vanishes far from the boundary,

$$\boxed{\text{grad}_{\mathbf{y}}\mathbf{u}^1 \xrightarrow{y_3 \to +\infty} \mathbf{0},} \tag{4.10}$$

---

[1] Regular boundaries are only considered, so that the projection is assumed to be unique for points $\mathbf{x}$ in the vicinity of Γ_t.

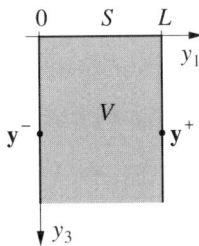

Fig. 4.3. Strip V – the unit cell for the boundary layer analysis.

so that the expansion (4.7) *matches* the macroscopic field far in the micro variable $\mathbf{y}$ from the boundary.

The asymptotic expansion for strain is obtained by taking the symmetrized gradient of the displacement field (4.7). When taking the spatial derivatives of the fields that depend on the two spatial variables, as for example $\mathbf{u}^1(\mathbf{x}', \mathbf{y})$ does, the following rule[2] is applied, cf. Sanchez-Palencia [125],

$$\mathrm{grad}(\cdot) = \mathrm{grad}_{\mathbf{x}}(\cdot) + \frac{1}{\varepsilon}\,\mathrm{grad}_{\mathbf{y}}(\cdot). \tag{4.11}$$

The asymptotic expansion of the strain field is thus

$$\mathbf{e}^\varepsilon(\mathbf{x}) = \mathbf{e}^0(\mathbf{x}', \mathbf{y}) + \varepsilon\mathbf{e}^1(\mathbf{x}', \mathbf{y}) + O(\varepsilon^2), \tag{4.12}$$

where the leading term $\mathbf{e}^0(\mathbf{x}', \mathbf{y})$ is given by

$$\boxed{\mathbf{e}^0(\mathbf{x}', \mathbf{y}) = \mathbf{e}_{\mathbf{x}}(\mathbf{u}^0) + \mathbf{e}_{\mathbf{y}}(\mathbf{u}^1),} \tag{4.13}$$

and $\mathbf{e}^1(\mathbf{x}', \mathbf{y}) = \mathbf{e}_{\mathbf{x}}(\mathbf{u}^1) + \mathbf{e}_{\mathbf{y}}(\mathbf{u}^2)$, where $\mathbf{u}^2(\mathbf{x}', \mathbf{y})$ is the second-order correction term in (4.7). Here, $\mathbf{e}_{\mathbf{x}}$ and $\mathbf{e}_{\mathbf{y}}$ denote the strains in, respectively, $\mathbf{x}$ and $\mathbf{y}$ variables,

$$\mathbf{e}_{\mathbf{x}}(\mathbf{u}) = \frac{1}{2}\,[\,\mathrm{grad}_{\mathbf{x}}\mathbf{u} + (\mathrm{grad}_{\mathbf{x}}\mathbf{u})^T\,], \quad \mathbf{e}_{\mathbf{y}}(\mathbf{u}) = \frac{1}{2}\,[\,\mathrm{grad}_{\mathbf{y}}\mathbf{u} + (\mathrm{grad}_{\mathbf{y}}\mathbf{u})^T\,]. \tag{4.14}$$

The leading term $\mathbf{e}^0$ in the expansion of the strain field is thus a sum of the macroscopic strain $\mathbf{e}_{\mathbf{x}}(\mathbf{u}^0)$ and a correction term $\mathbf{e}_{\mathbf{y}}(\mathbf{u}^1)$ associated with the displacement correction $\mathbf{u}^1$.

The asymptotic expansion of the stress follows from the constitutive equation $\boldsymbol{\sigma}^\varepsilon = \mathbf{L}\mathbf{e}^\varepsilon$, so that

$$\boldsymbol{\sigma}^\varepsilon(\mathbf{x}) = \boldsymbol{\sigma}^0(\mathbf{x}', \mathbf{y}) + \varepsilon\boldsymbol{\sigma}^1(\mathbf{x}', \mathbf{y}) + O(\varepsilon^2), \tag{4.15}$$

with

---

[2] Consider function $f(x)$ defined by $f(x) = g(x, y)$ with $y = x/\varepsilon$. The derivative of f is then given by $\mathrm{d}f/\mathrm{d}x = \partial g/\partial x + (1/\varepsilon)\,\partial g/\partial y$, cf. (4.11).

$$\boldsymbol{\sigma}^0(\mathbf{x}',\mathbf{y}) = \mathbf{L}\mathbf{e}^0(\mathbf{x}',\mathbf{y}), \qquad \boldsymbol{\sigma}^1(\mathbf{x}',\mathbf{y}) = \mathbf{L}\mathbf{e}^1(\mathbf{x}',\mathbf{y}). \tag{4.16}$$

Finally, the equilibrium equation is evaluated for $\boldsymbol{\sigma}^\varepsilon$, namely

$$\operatorname{div}\boldsymbol{\sigma}^\varepsilon = \varepsilon^{-1}\operatorname{div}_{\mathbf{y}}\boldsymbol{\sigma}^0 + (\operatorname{div}_{\mathbf{x}}\boldsymbol{\sigma}^0 + \operatorname{div}_{\mathbf{y}}\boldsymbol{\sigma}^1) + O(\varepsilon) = \mathbf{0}. \tag{4.17}$$

The micro-periodic prescribed traction $\mathbf{t}^{\mathrm{b}}$, cf. (4.3), can now be written in the form

$$\mathbf{t}^{\mathrm{b}}(\xi,\eta) = \bar{\mathbf{t}}^{\mathrm{b}}(\xi) + \mathbf{t}_{\mathrm{a}}^{\mathrm{b}}(\xi)\,\tilde{p}_\varepsilon(\eta), \tag{4.18}$$

where $\eta = \xi/\varepsilon$ is the microscopic counterpart to the macroscopic arc-length parameter ξ, and $\tilde{p}_\varepsilon(\eta) = \tilde{p}(\varepsilon\eta)$ is a periodic, purely fluctuating function with the period $L = l/\varepsilon$. The boundary condition (4.1)₃, expressed in terms of $\boldsymbol{\sigma}^\varepsilon$, becomes thus

$$\boldsymbol{\sigma}^0(\mathbf{x}',\mathbf{y})\mathbf{n} + \varepsilon\boldsymbol{\sigma}^1(\mathbf{x}',\mathbf{y})\mathbf{n} + O(\varepsilon^2) = \bar{\mathbf{t}}^{\mathrm{b}}(\xi(\mathbf{x}')) + \mathbf{t}_{\mathrm{a}}^{\mathrm{b}}(\xi(\mathbf{x}'))\,\tilde{p}_\varepsilon(\eta(\mathbf{y})), \tag{4.19}$$

where $\mathbf{y} \in S$ and S is the microscopic representation of the boundary Γ_{t}, cf. Fig. 4.3.

4.2.4 Microscopic Problem

The equations of the boundary layer, i.e. the *microscopic problem*, are now obtained by considering only the leading terms, i.e. the terms with the lowest exponent of ε, of the governing equations (equilibrium equation, constitutive equation, and boundary condition on Γ_{t}) evaluated in terms of $\mathbf{u}^\varepsilon$, $\mathbf{e}^\varepsilon$, and $\boldsymbol{\sigma}^\varepsilon$. The resulting boundary value problem is to find a V-periodic displacement correction term $\mathbf{u}^1(\mathbf{x}',\mathbf{y})$ that satisfies condition (4.10) and the following set of equations

$$\begin{cases} \operatorname{div}_{\mathbf{y}}\boldsymbol{\sigma}^0 = \mathbf{0}, & \mathbf{x}' \in \Gamma_{\mathrm{t}},\ \mathbf{y} \in V \\ \boldsymbol{\sigma}^0 = \mathbf{L}[\mathbf{e}_{\mathbf{x}}(\mathbf{u}^0) + \mathbf{e}_{\mathbf{y}}(\mathbf{u}^1)], & \mathbf{x}' \in \Gamma_{\mathrm{t}},\ \mathbf{y} \in V \\ \boldsymbol{\sigma}^0\mathbf{n} = \bar{\mathbf{t}}^{\mathrm{b}}(\xi(\mathbf{x}')) + \mathbf{t}_{\mathrm{a}}^{\mathrm{b}}(\xi(\mathbf{x}'))\,\tilde{p}_\varepsilon(\eta(\mathbf{y})), & \mathbf{x}' \in \Gamma_{\mathrm{t}},\ \mathbf{y} \in S \end{cases} \tag{4.20}$$

Regardless of the shape of domain Ω, the microscopic problem (4.20) is solved for a simple geometry, i.e. for a half-plane. Furthermore, due to V-periodicity of $\mathbf{u}^1$, it is sufficient to consider only the strip V as a periodic unit cell. At the same time, $\mathbf{x}' \in \Gamma_{\mathrm{t}}$ plays the role of parameter, and the macroscopic strain $\mathbf{e}_{\mathbf{x}}(\mathbf{u}^0)$, determined by the macroscopic problem (4.5), constitutes the input.

Remark 4.1. In view of the condition (4.10), the solution $\mathbf{u}^1$ is only defined up to an additive vector (rigid translation).

Remark 4.2. The microscopic problem (4.20) can be written in a simpler form by subtracting the constant macroscopic stress $\boldsymbol{\sigma}^{\mathrm{h}} = \mathbf{L}\mathbf{e}_{\mathbf{x}}(\mathbf{u}^0)$ from the microscopic stress $\boldsymbol{\sigma}^0$. However, here and in the subsequent sections, we formulate the microscopic problem in terms of the total microscopic stress $\boldsymbol{\sigma}^0$ because of its direct physical meaning in the case of contact and friction conditions studied in Sects. 4.4 and 4.5.

4.2.5 Average Stress in the Boundary Layer

Let us introduce the following averaging operation

$$\langle \varphi \rangle (\mathbf{x}', y_3) = \frac{1}{L} \int_0^L \varphi(\mathbf{x}', \mathbf{y}) \, dy_1, \tag{4.21}$$

which averages an arbitrary field $\varphi(\mathbf{x}', \mathbf{y})$ over the period $L = l/\varepsilon$ at a fixed distance y_3 from the boundary.

Consider now the average of the equilibrium equation $(4.20)_1$, namely

$$\mathbf{0} = \langle \mathrm{div}_{\mathbf{y}} \boldsymbol{\sigma}^0 \rangle = \mathrm{div}_{\mathbf{y}} \langle \boldsymbol{\sigma}^0 \rangle = \frac{d}{dy_3} (\langle \boldsymbol{\sigma}^0 \rangle \mathbf{n}), \tag{4.22}$$

where the last transformation is due to the fact that $\langle \boldsymbol{\sigma}^0 \rangle (\mathbf{x}', y_3)$ depends only on y_3. In view of (4.22), $(4.20)_3$, and $(4.5)_3$, we conclude that

$$\boxed{\langle \boldsymbol{\sigma}^0 \rangle \mathbf{n} = \boldsymbol{\sigma}^{\mathrm{h}} \mathbf{n} = \bar{\mathbf{t}}^{\mathrm{b}}(\xi) = \mathrm{const}(\xi),} \tag{4.23}$$

which also means that, at fixed $\mathbf{x}' \in \Gamma_{\mathrm{t}}$, the *exterior* part of the average stress $\langle \boldsymbol{\sigma}^0 \rangle (\mathbf{x}', y_3)$ is constant, i.e. it does not depend on y_3,

$$\langle \boldsymbol{\sigma}^0 \rangle_{\mathrm{A}} = \boldsymbol{\sigma}_{\mathrm{A}}^{\mathrm{h}} = \mathrm{const}(\xi). \tag{4.24}$$

Conditions (4.23) and (4.24) will prove useful in the case of contact boundary layers, considered below, and are also discussed in more detail in Chap. 5.

4.3 Micro-Periodic Prescribed Displacement

For completeness, let us briefly discuss the boundary value problem defined by (4.1) with a micro-periodic displacement $\mathbf{u}^{\mathrm{b}}$ prescribed on Γ_{u},

$$\mathbf{u}^{\mathrm{b}}(\xi) = \bar{\mathbf{u}}^{\mathrm{b}}(\xi) + \mathbf{u}_{\mathrm{a}}^{\mathrm{b}}(\xi)\tilde{p}(\xi), \tag{4.25}$$

where $\tilde{p}(\xi)$ is a l-periodic, purely oscillating function. Introducing the macro and micro spatial variables the prescribed displacement can be rewritten in the form

$$\mathbf{u}^{\mathrm{b}}(\xi, \eta) = \bar{\mathbf{u}}^{\mathrm{b}}(\xi) + \varepsilon \mathbf{u}_{\mathrm{a}}^{\mathrm{b}}(\xi)\tilde{p}_\varepsilon(\eta), \tag{4.26}$$

where, similarly to the case of micro-periodic surface tractions, $\xi \in (0, L)$ and $\eta = \xi/\varepsilon$ parameterize the boundary Γ_{u}, $\bar{\mathbf{u}}^{\mathrm{b}}(\xi)$ and $\mathbf{u}_{\mathrm{a}}^{\mathrm{b}}(\xi)$ are, respectively, the average and the amplitude, both slowly varying functions, and $\tilde{p}_\varepsilon(\eta) = (1/\varepsilon)\tilde{p}(\xi/\varepsilon)$ is a L-periodic, purely oscillating function. Note that, in order to keep the strains in the boundary layer bounded, the oscillating term in the prescribed displacement $\mathbf{u}^{\mathrm{b}}$ is now scaled by ε, cf. (4.26).

The *macroscopic problem* is obtained by prescribing *micro-homogeneous* displacement $\bar{\mathbf{u}}^{\mathrm{b}}$ on Γ_{u}, namely

$$\begin{cases} \operatorname{div} \boldsymbol{\sigma}^{\mathrm{h}} = \mathbf{0}, & \mathbf{x} \in \Omega \\ \boldsymbol{\sigma}^{\mathrm{h}} = \mathbf{L}\mathbf{e}(\mathbf{u}^0), & \mathbf{x} \in \Omega \\ \boldsymbol{\sigma}^{\mathrm{h}}\mathbf{n} = \mathbf{t}^{\mathrm{b}}, & \mathbf{x} \in \Gamma_{\mathrm{t}} \\ \mathbf{u}^0 = \bar{\mathbf{u}}^{\mathrm{b}}, & \mathbf{x} \in \Gamma_{\mathrm{u}} \end{cases} \qquad (4.27)$$

which is solved for the unknown macroscopic displacement field $\mathbf{u}^0(\mathbf{x})$. Again, the solution of this problem is valid everywhere within Ω, except in the vicinity of boundary Γ_{u}.

The asymptotic expansion of the displacement field in the vicinity of the boundary Γ_{u} is given by (4.7) with Γ_{t} and $\Gamma_{\mathrm{t}}^{!}$ replaced by Γ_{u} and $\Gamma_{\mathrm{u}}^{!}$, respectively. The correction term $\mathbf{u}^1(\mathbf{x}', \mathbf{y})$ is V-periodic, and its gradient vanishes far from the boundary, cf. (4.10). However, in the present case, the boundary condition on Γ_{u}, $\mathbf{u}^{\varepsilon} = \mathbf{u}^{\mathrm{b}}$, uniquely defines the correction term along the boundary,

$$\mathbf{u}^1(\mathbf{x}', \mathbf{y}) = \mathbf{u}_{\mathrm{a}}^{\mathrm{b}}(\xi(\mathbf{x}')) \, \tilde{p}_{\varepsilon}(\eta(\mathbf{y})), \qquad \mathbf{x}' \in \Gamma_{\mathrm{u}}, \ \mathbf{y} \in S \qquad (4.28)$$

in contrast to the case of micro-inhomogeneous tractions, where the correction term is defined up to an additive vector, cf. Remark 4.1.

Following the procedure outlined in Sect. 4.2, the following *microscopic problem* is obtained: find a V-periodic displacement correction term $\mathbf{u}^1(\mathbf{x}', \mathbf{y})$ that satisfies condition (4.10) and the set of equations

$$\begin{cases} \operatorname{div}_{\mathbf{y}} \boldsymbol{\sigma}^0 = \mathbf{0}, & \mathbf{x}' \in \Gamma_{\mathrm{u}}, \ \mathbf{y} \in V \\ \boldsymbol{\sigma}^0 = \mathbf{L}[\mathbf{e}_{\mathbf{x}}(\mathbf{u}^0) + \mathbf{e}_{\mathbf{y}}(\mathbf{u}^1)], & \mathbf{x}' \in \Gamma_{\mathrm{u}}, \ \mathbf{y} \in V \\ \mathbf{u}^1(\mathbf{x}', \mathbf{y}) = \mathbf{u}_{\mathrm{a}}^{\mathrm{b}}(\xi(\mathbf{x}')) \, \tilde{p}_{\varepsilon}(\eta(\mathbf{y})), & \mathbf{x}' \in \Gamma_{\mathrm{u}}, \ \mathbf{y} \in S \end{cases} \qquad (4.29)$$

Again, this problem is parameterized by $\mathbf{x}' \in \Gamma_{\mathrm{u}}$ and $\mathbf{e}_{\mathbf{x}}(\mathbf{u}^0)$ is the input known from the macroscopic problem (4.27).

4.4 Frictionless Contact with a Rigid Obstacle

4.4.1 Problem Statement

Consider now the problem of an elastic body in contact with a rigid obstacle, Fig. 4.4(a). The body occupies domain Ω_{r}, and the potential contact surface Γ_{r} is assumed to be rough (i.e. micro-undulated). We will also consider the nominal (smooth) contact surface Γ_{n} which is a part of the boundary of the corresponding smooth domain Ω_{n}, cf. Fig. 4.4(b). In the present context, "smoothness" refers to the lack of roughness, while a rough (i.e. non-smooth) surface may be smooth in the mathematical sense (i.e. continuous and differentiable).

Let the equation of the nominal contact surface Γ_{n} be denoted by $\hat{\mathbf{x}}_{\mathrm{n}}(\xi)$,

$$\Gamma_{\mathrm{n}} = \{\mathbf{x}: \ \mathbf{x} = \hat{\mathbf{x}}_{\mathrm{n}}(\xi); \ \xi \in (0, L)\}, \qquad (4.30)$$

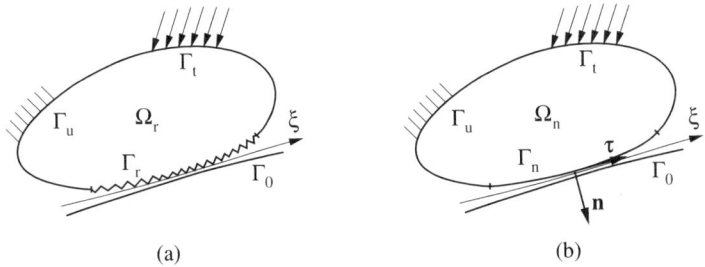

Fig. 4.4. Contact of a rough body Ω_r with a smooth and rigid obstacle Γ_0: (a) rough contact surface Γ_r and (b) smooth nominal contact surface Γ_n with the corresponding domain Ω_n.

and the equation of the actual rough contact surface Γ_r by $\hat{\mathbf{x}}_r(\xi)$,

$$\Gamma_r = \{\mathbf{x}: \ \mathbf{x} = \hat{\mathbf{x}}_r(\xi); \ \xi \in (0, L)\}, \quad \hat{\mathbf{x}}_r(\xi) = \hat{\mathbf{x}}_n(\xi) - \tilde{p}(\xi)\,\mathbf{n}(\xi), \qquad (4.31)$$

where $\mathbf{n}(\xi)$ is the unit outward normal to Γ_n and

$$\tilde{p}(\xi) \geq 0, \qquad \tilde{p}(\xi) = \tilde{p}(\xi + l), \qquad P_0 = \{\xi: \ \tilde{p}(\xi) = 0\} \neq \emptyset. \qquad (4.32)$$

Here, $\tilde{p}(\xi)$ is a non-negative, l-periodic function, and P_0 is a non-empty set of points ξ at which $\tilde{p}(\xi) = 0$. Thus, in the present setting, the nominal surface Γ_n is an outer envelope of the rough surface Γ_r, and $\Omega_r \subset \Omega_n$. Finally, the rigid obstacle is represented by surface Γ_0,

$$\Gamma_0 = \{\mathbf{x}: \ \mathbf{x} = \hat{\mathbf{x}}_0(\xi); \ \xi \in (0, L)\}, \qquad \hat{\mathbf{x}}_0(\xi) = \hat{\mathbf{x}}_n(\xi) + g_0(\xi)\,\mathbf{n}(\xi), \qquad (4.33)$$

where $g_0(\xi)$ is the initial normal gap between the obstacle Γ_0 and the nominal contact surface Γ_n.

We assume that the displacements are small. Accordingly, in the potential contact zone, the difference between the unit vectors normal to surfaces Γ_r, Γ_n, and Γ_0 is neglected. The normal gap g_N in the deformed state is thus defined by

$$g_N(\xi) = [\hat{\mathbf{x}}_0(\xi) - \hat{\mathbf{x}}_r(\xi) - \mathbf{u}(\hat{\mathbf{x}}_r(\xi))] \cdot \mathbf{n}(\xi)$$
$$= g_0(\xi) + \tilde{p}(\xi) - \mathbf{u}(\hat{\mathbf{x}}_r(\xi)) \cdot \mathbf{n}(\xi), \qquad (4.34)$$

and the normal contact traction $t_N(\xi)$ is defined by

$$t_N(\xi) = \mathbf{n}(\xi) \cdot \boldsymbol{\sigma}(\hat{\mathbf{x}}_r(\xi))\mathbf{n}(\xi). \qquad (4.35)$$

In the case of frictionless contact, we have additionally $t_T = \boldsymbol{\tau} \cdot \boldsymbol{\sigma}\mathbf{n} = 0$ and $\boldsymbol{\sigma}\mathbf{n} = t_N\mathbf{n}$, where $\boldsymbol{\tau}$ is the unit vector tangential to the nominal contact surface Γ_n.

The normal gap g_N and the contact traction t_N are related by the unilateral contact (impenetrability) condition which can be written in the form of the classical Signorini condition

$$g_N \geq 0, \qquad t_N \leq 0, \qquad g_N t_N = 0. \tag{4.36}$$

Accordingly, three contact states can be distinguished:

i. separation: $g_N > 0$, $t_N = 0$,
ii. contact: $g_N = 0$, $t_N < 0$,
iii. grazing contact: $g_N = 0$, $t_N = 0$.

The boundary value problem of frictionless contact of the rough body, the *background problem*, is to find displacement field $\mathbf{u}(\mathbf{x})$, $\mathbf{x} \in \Omega_r$, satisfying

$$\begin{cases} \operatorname{div} \boldsymbol{\sigma} = \mathbf{0}, & \mathbf{x} \in \Omega_r \\ \boldsymbol{\sigma} = \mathbf{Le}(\mathbf{u}), & \mathbf{x} \in \Omega_r \\ \boldsymbol{\sigma}\mathbf{n} = \mathbf{t}^b, & \mathbf{x} \in \Gamma_t \\ \mathbf{u} = \mathbf{u}^b, & \mathbf{x} \in \Gamma_u \\ \boldsymbol{\sigma}\mathbf{n} = t_N \mathbf{n} + t_T \boldsymbol{\tau}, & \mathbf{x} \in \Gamma_r \\ g_N \geq 0, \ t_N \leq 0, \ g_N t_N = 0, & \mathbf{x} \in \Gamma_r \\ t_T = 0, & \mathbf{x} \in \Gamma_r \end{cases} \tag{4.37}$$

4.4.2 Macroscopic Problem

The *macroscopic problem* is obtained by replacing the rough surface Γ_r with the smooth nominal surface Γ_n. The problem is defined on the corresponding (smooth) domain Ω_n, namely

$$\begin{cases} \operatorname{div} \boldsymbol{\sigma}^h = \mathbf{0}, & \mathbf{x} \in \Omega_n \\ \boldsymbol{\sigma}^h = \mathbf{Le}(\mathbf{u}^0), & \mathbf{x} \in \Omega_n \\ \boldsymbol{\sigma}^h\mathbf{n} = \mathbf{t}^b, & \mathbf{x} \in \Gamma_t \\ \mathbf{u}^0 = \mathbf{u}^b, & \mathbf{x} \in \Gamma_u \\ \boldsymbol{\sigma}^h\mathbf{n} = t_N^h \mathbf{n} + t_T^h \boldsymbol{\tau}, & \mathbf{x} \in \Gamma_n \\ g_N^h \geq 0, \ t_N^h \leq 0, \ g_N^h t_N^h = 0, & \mathbf{x} \in \Gamma_n \\ t_T^h = 0, & \mathbf{x} \in \Gamma_n \end{cases} \tag{4.38}$$

where $\mathbf{u}^0(\mathbf{x})$, $\mathbf{x} \in \Omega_n$, is the unknown macroscopic displacement. The components of the macroscopic contact traction $\mathbf{t}^h = t_N^h \mathbf{n} + t_T^h \boldsymbol{\tau}$ are given by

$$t_N^h = \mathbf{n} \cdot \boldsymbol{\sigma}^h \mathbf{n}, \qquad t_T^h = \boldsymbol{\tau} \cdot \boldsymbol{\sigma}^h \mathbf{n} \tag{4.39}$$

Finally, the macroscopic normal gap g_N^h is defined by

$$\begin{aligned} g_N^h(\xi) &= [\hat{\mathbf{x}}_0(\xi) - \hat{\mathbf{x}}_n(\xi) - \mathbf{u}^0(\hat{\mathbf{x}}_n(\xi))] \cdot \mathbf{n}(\xi) \\ &= g_0(\xi) - \mathbf{u}^0(\hat{\mathbf{x}}_n(\xi)) \cdot \mathbf{n}(\xi). \end{aligned} \tag{4.40}$$

4.4.3 Asymptotic Expansions

Introducing the small parameter $\varepsilon = l/L \ll 1$, and the micro variable $\eta = \xi/\varepsilon$, the function describing the micro-periodically undulated contact surface Γ_{r} can be rewritten in the form

$$\hat{\mathbf{x}}_{\mathrm{r}}(\xi,\eta) = \hat{\mathbf{x}}_{\mathrm{n}}(\xi) - \varepsilon \tilde{p}_\varepsilon(\eta)\,\mathbf{n}(\xi), \qquad \tilde{p}_\varepsilon(\eta) = \tilde{p}(\varepsilon\eta)/\varepsilon. \qquad (4.41)$$

According to $(4.41)_2$, the roughness is scaled with ε homothetically, i.e. the ratio of asperity height and spacing is preserved.

The asymptotic expansion of the displacement field in the vicinity of the boundary Γ_{r} is next introduced,

$$\mathbf{u}^\varepsilon(\mathbf{x}) = \mathbf{u}^0(\mathbf{x}) + \varepsilon \mathbf{u}^1(\mathbf{x}',\mathbf{y}) + O(\varepsilon^2), \quad \mathbf{x} \in \Gamma_{\mathrm{r}}^+, \ \mathbf{x}' \in \Gamma_{\mathrm{n}}, \ \mathbf{y} \in V, \qquad (4.42)$$

where $\mathbf{x}'$ is the orthogonal projection of $\mathbf{x}$ onto the nominal contact surface Γ_{n}, and Γ_{r}^+ denotes the vicinity of Γ_{r}. Although the domains of the background problem (4.37) and of the macroscopic problem (4.38), Ω_{r} and Ω_{n}, respectively, are different, the expansion (4.42) is well defined for all $\mathbf{x} \in \Gamma_{\mathrm{r}}^+$, because $\Omega_{\mathrm{r}} \subset \Omega_{\mathrm{n}}$.

The correction term $\mathbf{u}^1(\mathbf{x}',\mathbf{y})$ is V-periodic, and its gradient vanishes far from the surface, cf. (4.10). The strip V, see Fig. 4.5, is now defined by

$$V = \{(y_1,y_3):\ y_1 = \eta;\ y_3 \geq \tilde{p}_\varepsilon(\eta);\ \eta \in (0,L)\}, \qquad (4.43)$$

and the undulated surface S_{r}, i.e. the contact surface at the micro-scale, is given by

$$S_{\mathrm{r}} = \{(y_1,y_3):\ y_1 = \eta;\ y_3 = \tilde{p}_\varepsilon(\eta);\ \eta \in (0,L)\}. \qquad (4.44)$$

The asymptotic expansion of the normal gap function $g_{\mathrm{N}}(\xi)$ is obtained by combining (4.42) and (4.34), namely

$$g_{\mathrm{N}}^\varepsilon(\xi) = g_{\mathrm{N}}^0(\xi) + \varepsilon g_{\mathrm{N}}^1(\xi,\eta) + O(\varepsilon^2), \qquad (4.45)$$

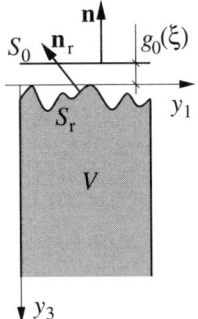

Fig. 4.5. Strip V – the unit cell of the rough boundary layer.

where

$$g_N^0(\xi) = g_0(\xi) - \mathbf{u}^0(\hat{\mathbf{x}}_n(\xi)) \cdot \mathbf{n}(\xi), \tag{4.46}$$

$$g_N^1(\xi, \eta) = \tilde{p}_\varepsilon(\eta) - \mathbf{u}^1(\hat{\mathbf{x}}_n(\xi), \hat{\mathbf{y}}_r(\eta)) \cdot \mathbf{n}(\xi), \tag{4.47}$$

and $\hat{\mathbf{y}}_r(\eta)$ is the parametric representation of surface S_r. Comparing (4.46) and (4.40), we notice that

$$\boxed{g_N^0(\xi) = g_N^h(\xi).} \tag{4.48}$$

In fact, $g_N^0(\xi)$ is the gap between the nominal microscopic contact surface S_n and plane S_0, parallel to S_n, which represents the rigid obstacle Γ_0 at the micro scale, cf. Fig. 4.5.

The expansion of the contact traction follows from the expansion of the stress (4.15), namely

$$t_N^\varepsilon(\xi) = t_N^0(\xi, \eta) + \varepsilon t_N^1(\xi, \eta) + O(\varepsilon^2), \tag{4.49}$$

where

$$t_N^0(\xi, \eta) = \mathbf{n}(\xi) \cdot \boldsymbol{\sigma}^0(\hat{\mathbf{x}}_n(\xi), \hat{\mathbf{y}}_r(\eta))\mathbf{n}(\xi). \tag{4.50}$$

In view of the impenetrability condition $(4.38)_6$ at the macro scale, the potential contact surface Γ_n is divided into the *macroscopic contact zone* (with $t_N^h < 0$) and the *macroscopic separation zone* (with $g_N^h > 0$). Below, it is shown that the boundary layer equations are different within these two zones. It is also shown that, in the present context, the points of macroscopic grazing contact can be included into the separation zone.

4.4.4 Macroscopic Contact Zone

In the macroscopic contact zone, the macroscopic normal gap is equal to zero, $g_N^h = 0$, and in view of (4.45)–(4.48) we have

$$g_N^\varepsilon(\xi) = \varepsilon g_N^1(\xi, \eta) + O(\varepsilon^2), \tag{4.51}$$

so that now g_N^1 constitutes the leading term of g_N^ε. The *microscopic unilateral contact condition* takes thus the form

$$g_N^1(\xi, \eta) \geq 0, \qquad t_N^0(\xi, \eta) \leq 0, \qquad g_N^1(\xi, \eta)\, t_N^0(\xi, \eta) = 0, \tag{4.52}$$

with g_N^1 and t_N^0 given by (4.47) and (4.50), respectively.

The *microscopic problem* is thus specified by the following set of equations

$$\begin{cases} \operatorname{div}_\mathbf{y} \boldsymbol{\sigma}^0 = \mathbf{0}, & \mathbf{x}' \in \Gamma_n,\ \mathbf{y} \in V \\ \boldsymbol{\sigma}^0 = \mathbf{L}[\mathbf{e}_\mathbf{x}(\mathbf{u}^0) + \mathbf{e}_\mathbf{y}(\mathbf{u}^1)], & \mathbf{x}' \in \Gamma_n,\ \mathbf{y} \in V \\ \boldsymbol{\sigma}^0 \mathbf{n} = t_N^0 \mathbf{n}, & \mathbf{x}' \in \Gamma_n,\ \mathbf{y} \in S_r \\ g_N^1 \geq 0,\ t_N^0 \leq 0,\ g_N^1 t_N^0 = 0, & \mathbf{x}' \in \Gamma_n,\ \mathbf{y} \in S_r \end{cases} \tag{4.53}$$

with unknown displacement $\mathbf{u}^1(\mathbf{x}', \mathbf{y})$ which is a V-periodic correction term satisfying (4.10). In the case of micro-inhomogeneous traction, cf. Sect. 4.2, the displacement correction term $\mathbf{u}^1$ is only defined up to an additive vector. In the present case, the normal component of $\mathbf{u}^1$, which affects the normal gap g_N^1, is constrained by the requirement that the average contact pressure at the micro-scale is equal to the macroscopic contact pressure,

$$\langle t_N^0 \rangle(\xi) = t_N^h(\xi). \tag{4.54}$$

As a result, the solution $\mathbf{u}^1$ is defined up to an additive vector *tangential* to the nominal contact surface, as the contact condition $(4.53)_4$ is not sensitive to tangential displacements, cf. (4.47).

Note that constraint (4.54) need not be enforced separately, since, from (4.10) and $(4.53)_2$, it follows that $\boldsymbol{\sigma}^0(\mathbf{x}', \mathbf{y}) \to \boldsymbol{\sigma}^h(\mathbf{x}')$ as $y_3 \to \infty$, and then condition (4.54) follows from (4.23) and (4.39).

Solution of the microscopic problem (4.53) divides the microscopic contact surface S_r into microscopic separation and contact zones. In contact mechanics, the latter is usually referred to as the *real area of contact*, and is characterized by the real contact area fraction α,

$$\alpha(\xi) = \frac{1}{L} \int_0^L I(g_N^1(\xi, \eta)) \, \mathrm{d}\eta, \qquad I(g_N^1) = \begin{cases} 1 \text{ if } g_N^1 = 0, \\ 0 \text{ if } g_N^1 > 0, \end{cases} \tag{4.55}$$

i.e. the ratio of the area of microscopic contact zones to the area of the nominal contact surface S_n. By definition $0 \le \alpha \le 1$.

4.4.5 Macroscopic Separation Zone

In the macroscopic separation zone, the macroscopic gap is greater than zero, $g_N^h = g_N^0 > 0$. This implies that for $\varepsilon \to 0$ separation occurs also at the micro scale, cf. (4.45), so that the unilateral contact conditions are automatically satisfied.

Accordingly, the *microscopic problem* is a purely elastic problem of equilibrium of strip V with zero surface traction prescribed on S_r. The problem is thus to find a V-periodic displacement $\mathbf{u}^1(\mathbf{x}', \mathbf{y})$ satisfying (4.10) and

$$\begin{cases} \mathrm{div}_\mathbf{y} \boldsymbol{\sigma}^0 = \mathbf{0}, & \mathbf{x}' \in \Gamma_n, \, \mathbf{y} \in V \\ \boldsymbol{\sigma}^0 = \mathbf{L}[\mathbf{e}_\mathbf{x}(\mathbf{u}^0) + \mathbf{e}_\mathbf{y}(\mathbf{u}^1)], & \mathbf{x}' \in \Gamma_n, \, \mathbf{y} \in V \\ \boldsymbol{\sigma}^0 \mathbf{n} = \mathbf{0}, & \mathbf{x}' \in \Gamma_n, \, \mathbf{y} \in S_r \end{cases} \tag{4.56}$$

where, consistently with the adopted assumptions, the difference between $\mathbf{n}_r$, the local normal to S_r, and $\mathbf{n}$ has been neglected. The solution, $\mathbf{u}^1$, is only defined up to an additive vector, as in the case of micro-inhomogeneous surface traction, Sect. 4.2.

From the point of view of the present boundary layer analysis, the points of macroscopic grazing contact (i.e. such that $g_N^h = 0$ and $t_N^h = 0$) do not require separate treatment, and the boundary layer equations (4.56) of the macroscopic separation zone apply also in that case. Indeed, in view of the no-tension condition, $t_N^0(\xi, \eta) \leq 0$, condition of vanishing macroscopic contact traction, $t_N^h(\xi) = 0$, implies that the microscopic contact traction vanishes along S_r, i.e. $t_N^0(\xi, \eta) = 0$ for all η, and this implies boundary condition (4.56)$_3$.

4.5 Frictional Contact

4.5.1 Problem Statement

Let us finally consider the case of frictional contact. Since friction is a path-dependent phenomenon, the displacement must be considered as a function of time t. The discussion below is restricted to the case of *rate-independent* friction laws, so that t stands for any time-like parameter, e.g. the load multiplier. Moreover, the changes in time are assumed sufficiently slow for the inertia forces to be neglected (quasi-static problem).

The friction law relates the sliding velocity $\dot{g}_T$ and the friction stress t_T. The sliding velocity $\dot{g}_T$ is the tangential component of the relative velocity at the contact point,

$$\dot{g}_T(\xi, t) = -\dot{\mathbf{u}}(\hat{\mathbf{x}}_r(\xi), t) \cdot \boldsymbol{\tau}(\xi), \qquad (4.57)$$

where $\dot{\mathbf{u}}(\mathbf{x}, t) = \partial \mathbf{u} / \partial t$ is the velocity of a point $\mathbf{x} \in \Omega_r$, and $\boldsymbol{\tau}$ is the unit vector tangential to the nominal contact surface Γ_n. For simplicity, the velocity of the obstacle is assumed to be equal to zero. The friction stress t_T is the tangential component of the contact traction vector $\mathbf{t}$, so that

$$\mathbf{t} = \boldsymbol{\sigma}\mathbf{n} = t_N\mathbf{n} + t_T\boldsymbol{\tau}, \qquad t_T = \boldsymbol{\tau} \cdot \boldsymbol{\sigma}\mathbf{n}. \qquad (4.58)$$

Both definitions (4.57) and (4.58) rely on the assumption that the displacements are small.

The boundary value problem of quasi-static frictional contact is to find the displacement field $\mathbf{u}(\mathbf{x}, t)$, $\mathbf{x} \in \Omega_r$, $t \in (0, T)$, satisfying equations of the frictionless contact problem (4.37) with the condition $t_T = 0$, cf. (4.37)$_7$, replaced by the following friction law

$$\left. \begin{array}{ll} |t_T| \leq \tau_s & \text{if } g_N = 0, \ \dot{g}_T = 0 \\ t_T = \tau_s \, \text{sign}(\dot{g}_T) & \text{if } g_N = 0, \ \dot{g}_T \neq 0 \\ t_T = 0 & \text{if } g_N > 0 \end{array} \right\} \quad \text{on } \Gamma_r \qquad (4.59)$$

where τ_s is the threshold friction stress.

Since the friction law relates the friction stress and the velocity $\dot{\mathbf{u}}$, cf. (4.57), the resulting boundary value problem is, in fact, a rate problem. Thus, given the solution in the time interval $(0, t)$, the rate $\dot{\mathbf{u}}(\mathbf{x}, t)$ is to be found by solving the rate form of the governing equations (4.37) and (4.59). The detailed form of the rate equations is omitted here, and can be found, for example, in Klarbring [62].

To fix the attention, we assume that the threshold friction stress τ_s is the sum of the classical Coulomb friction component, with the friction coefficient $\mu \geq 0$, and the adhesive friction component $\tau_a \geq 0$, namely

$$\tau_s = \tau_a + \mu |t_N| = \tau_a - \mu t_N \geq 0. \tag{4.60}$$

A purely adhesive friction law is obtained for $\mu = 0$, and the Coulomb law is recovered for $\tau_a = 0$. The Coulomb-like component may be regarded as a representation of frictional interactions at the scale smaller than the microscale in the present analysis.

4.5.2 Macroscopic Problem

As in the frictionless case, the *macroscopic problem* corresponds to the nominal contact surface Γ_n and to the associated domain Ω_n. The unknown is the displacement field $\mathbf{u}^0 = \mathbf{u}^0(\mathbf{x}, t)$, $\mathbf{x} \in \Omega_n$, $t \in (0, T)$, which satisfies (4.38) with condition $(4.38)_7$ replaced by the macroscopic friction law

$$\left. \begin{array}{ll} |t_T^h| \leq \tau_s^h & \text{if } g_N^h = 0,\ \dot{g}_T^h = 0 \\ t_T^h = \tau_s^h \, \text{sign}(\dot{g}_T^h) & \text{if } g_N^h = 0,\ \dot{g}_T^h \neq 0 \\ t_T^h = 0 & \text{if } g_N^h > 0 \end{array} \right\} \quad \text{on } \Gamma_n, \tag{4.61}$$

where τ_s^h is the macroscopic threshold friction stress. The macroscopic sliding velocity and friction stress derive from (4.57) and (4.58), namely

$$\dot{g}_T^h(\xi, t) = -\dot{\mathbf{u}}^0(\hat{\mathbf{x}}_n(\xi), t) \cdot \boldsymbol{\tau}(\xi), \tag{4.62}$$

and

$$t_T^h(\xi, t) = \boldsymbol{\tau}(\xi) \cdot \boldsymbol{\sigma}^h(\hat{\mathbf{x}}_n(\xi), t) \mathbf{n}(\xi). \tag{4.63}$$

Because of the friction law (4.61), the macroscopic problem is a rate boundary value problem.

Let us anticipate a result obtained in Sects. 4.5.4 and 4.5.5 below. Due to the particular form (4.60) of the friction law, the macroscopic threshold friction stress τ_s^h depends on the solution of the microscopic problem, and is given by

$$\tau_s^h = \alpha \tau_a - \mu t_N^h, \tag{4.64}$$

where α is the fraction of the real contact area, cf. (4.55). As a result, the macroscopic problem is coupled with the microscopic one, so that both problems must be solved in parallel. This is in contrast to the cases discussed in the previous sections, where the macroscopic problem can be solved by assuming homogeneous boundary or contact conditions without referring to the solution in the boundary layer.

4.5.3 Asymptotic Expansions

Accounting for the time-dependence, the asymptotic expansion of the displacement field in the vicinity of Γ_r is assumed in the form

$$\mathbf{u}^\varepsilon(\mathbf{x}, t) = \mathbf{u}^0(\mathbf{x}, t) + \varepsilon \mathbf{u}^1(\mathbf{x}', \mathbf{y}, t) + O(\varepsilon^2), \quad \mathbf{x} \in \Gamma_r^+, \ \mathbf{x}' \in \Gamma_n, \ \mathbf{y} \in V, \quad (4.65)$$

where the correction term $\mathbf{u}^1$ is defined on the strip V. This implies the following expansion of the velocity

$$\dot{\mathbf{u}}^\varepsilon(\mathbf{x}, t) = \dot{\mathbf{u}}^0(\mathbf{x}, t) + \varepsilon \dot{\mathbf{u}}^1(\mathbf{x}', \mathbf{y}, t) + O(\varepsilon^2), \quad (4.66)$$

and from (4.57) the expansion of the sliding velocity is

$$\dot{g}_T^\varepsilon(\xi, t) = \dot{g}_T^0(\xi, t) + \varepsilon \dot{g}_T^1(\xi, \eta, t) + O(\varepsilon^2), \quad (4.67)$$

where

$$\dot{g}_T^0(\xi) = -\mathbf{u}^0(\hat{\mathbf{x}}_n(\xi)) \cdot \boldsymbol{\tau}(\xi), \quad (4.68)$$

$$\dot{g}_T^1(\xi, \eta) = -\mathbf{u}^1(\hat{\mathbf{x}}_n(\xi), \hat{\mathbf{y}}_r(\eta)) \cdot \boldsymbol{\tau}(\xi). \quad (4.69)$$

As in the case of the normal gap, cf. (4.48), we have

$$\boxed{\dot{g}_T^0(\xi) = \dot{g}_T^h(\xi).} \quad (4.70)$$

The expansion of the friction stress follows from (4.15) and (4.58), viz.

$$t_T^\varepsilon(\xi) = t_T^0(\xi, \eta) + \varepsilon t_T^1(\xi, \eta) + O(\varepsilon^2), \quad (4.71)$$

where

$$t_T^0(\xi, \eta) = \boldsymbol{\tau}(\xi) \cdot \boldsymbol{\sigma}^0(\hat{\mathbf{x}}_n(\xi), \hat{\mathbf{y}}_r(\eta))\mathbf{n}(\xi). \quad (4.72)$$

As discussed in Sect. 4.4, the potential nominal contact surface Γ_n is divided into macroscopic contact and separation zones. In the case of frictional contact, the macroscopic contact zone is further divided into macroscopic sliding and sticking zones. The corresponding boundary layer equations are discussed below. The macroscopic separation zone does not require separate analysis, as the discussion of Sect. 4.4.5 fully applies also in the present case.

4.5.4 Macroscopic Sliding Zone

The *macroscopic sliding zone* is a part of the macroscopic contact zone with non-zero sliding velocity, $\dot{g}_T^h \neq 0$. Thus, in view of (4.67) and (4.70), in the macroscopic sliding zone, we have $\dot{g}_T^\varepsilon \neq 0$ for $\varepsilon \to 0$, and sliding occurs at all the points of the microscopic contact zone.

The *microscopic problem* is thus obtained by combining the boundary layer equations (4.53) of the frictionless case with the condition of *sliding* friction, namely

$$\begin{cases} \text{div}_\mathbf{y}\boldsymbol{\sigma}^0 = \mathbf{0}, & \mathbf{x}' \in \Gamma_\mathrm{n}, \ \mathbf{y} \in V \\ \boldsymbol{\sigma}^0 = \mathbf{L}[\mathbf{e}_\mathbf{x}(\mathbf{u}^0) + \mathbf{e}_\mathbf{y}(\mathbf{u}^1)], & \mathbf{x}' \in \Gamma_\mathrm{n}, \ \mathbf{y} \in V \\ \boldsymbol{\sigma}^0\mathbf{n} = t_\mathrm{N}^0\mathbf{n} + t_\mathrm{T}^0\boldsymbol{\tau}, & \mathbf{x}' \in \Gamma_\mathrm{n}, \ \mathbf{y} \in S_\mathrm{r} \\ g_\mathrm{N}^1 \geq 0, \ t_\mathrm{N}^0 \leq 0, \ g_\mathrm{N}^1 t_\mathrm{N}^0 = 0, & \mathbf{x}' \in \Gamma_\mathrm{n}, \ \mathbf{y} \in S_\mathrm{r} \\ t_\mathrm{T}^0 = \begin{cases} \tau_\mathrm{s}^0 \, \text{sign}(\dot{g}_\mathrm{T}^\mathrm{h}) \text{ if } g_\mathrm{N}^1 = 0 \\ 0 \qquad\qquad\quad \text{if } g_\mathrm{N}^1 > 0 \end{cases} & \mathbf{x}' \in \Gamma_\mathrm{n}, \ \mathbf{y} \in S_\mathrm{r} \end{cases} \tag{4.73}$$

where

$$\tau_\mathrm{s}^0(\xi, \eta) = \tau_\mathrm{a} - \mu t_\mathrm{N}^0(\xi, \eta). \tag{4.74}$$

The unknown in (4.73) is a V-periodic displacement correction $\mathbf{u}^1(\mathbf{x}', \mathbf{y})$ satisfying (4.10). Since the sliding velocity is determined by the macroscopic problem, the boundary value problem (4.73) is *not* a rate problem. As in the frictionless case, the normal component of $\mathbf{u}^1$ is constrained by the unilateral contact condition (4.73)$_4$, while neither the unilateral contact condition nor the friction condition (4.73)$_5$ are sensitive to the tangential component of $\mathbf{u}^1$. Thus, the solution $\mathbf{u}^1$ is defined up to an additive vector tangential to the nominal contact surface.

By an argument similar to that justifying (4.54) we have

$$\langle t_\mathrm{N}^0 \rangle(\xi) = t_\mathrm{N}^\mathrm{h}(\xi), \qquad \langle t_\mathrm{T}^0 \rangle(\xi) = t_\mathrm{T}^\mathrm{h}(\xi), \tag{4.75}$$

and the macroscopic threshold friction stress $\tau_\mathrm{s}^\mathrm{h}$ can be obtained from (4.74) and (4.75), namely

$$t_\mathrm{T}^\mathrm{h} = \langle t_\mathrm{T}^0 \rangle = (\alpha\tau_\mathrm{a} - \mu t_\mathrm{N}^\mathrm{h}) \, \text{sign}(\dot{g}_\mathrm{T}^\mathrm{h}) = \tau_\mathrm{s}^\mathrm{h} \, \text{sign}(\dot{g}_\mathrm{T}^\mathrm{h}). \tag{4.76}$$

The macroscopic friction law (4.61) and the expression (4.64) for the macroscopic friction stress $\tau_\mathrm{s}^\mathrm{h}$ are thus verified in the case of macroscopic sliding, $\dot{g}_\mathrm{T}^\mathrm{h} \neq 0$.

4.5.5 Macroscopic Sticking Zone

The *macroscopic sticking zone* is characterized by $\dot{g}_\mathrm{T}^\mathrm{h} = \dot{g}_\mathrm{T}^0 = 0$, so that

$$\dot{g}_\mathrm{T}^\varepsilon(\xi, t) = \varepsilon \dot{g}_\mathrm{T}^1(\xi, \eta, t) + O(\varepsilon^2), \tag{4.77}$$

and $\dot{g}_\mathrm{T}^1$ is now the leading term of $\dot{g}_\mathrm{T}^\varepsilon$.

The *microscopic problem* is thus specified by the following equations

$$\begin{cases} \text{div}_\mathbf{y}\boldsymbol{\sigma}^0 = \mathbf{0}, & \mathbf{x}' \in \Gamma_\mathrm{n}, \ \mathbf{y} \in V \\ \boldsymbol{\sigma}^0 = \mathbf{L}[\mathbf{e}_\mathbf{x}(\mathbf{u}^0) + \mathbf{e}_\mathbf{y}(\mathbf{u}^1)], & \mathbf{x}' \in \Gamma_\mathrm{n}, \ \mathbf{y} \in V \\ \boldsymbol{\sigma}^0\mathbf{n} = t_\mathrm{N}^0\mathbf{n} + t_\mathrm{T}^0\boldsymbol{\tau}, & \mathbf{x}' \in \Gamma_\mathrm{n}, \ \mathbf{y} \in S_\mathrm{r} \\ g_\mathrm{N}^1 \geq 0, \ t_\mathrm{N}^0 \leq 0, \ g_\mathrm{N}^1 t_\mathrm{N}^0 = 0, & \mathbf{x}' \in \Gamma_\mathrm{n}, \ \mathbf{y} \in S_\mathrm{r} \\ \begin{cases} |t_\mathrm{T}^0| \leq \tau_\mathrm{s}^0 & \text{if } g_\mathrm{N}^1 = 0, \ \dot{g}_\mathrm{T}^1 = 0 \\ t_\mathrm{T}^0 = \tau_\mathrm{s}^0 \, \text{sign}(\dot{g}_\mathrm{T}^1) & \text{if } g_\mathrm{N}^1 = 0, \ \dot{g}_\mathrm{T}^1 \neq 0 \\ t_\mathrm{T}^0 = 0 & \text{if } g_\mathrm{N}^1 > 0 \end{cases} & \mathbf{x}' \in \Gamma_\mathrm{n}, \ \mathbf{y} \in S_\mathrm{r} \end{cases} \tag{4.78}$$

with $\mathbf{u}^1(\mathbf{x}', \mathbf{y}, t)$ as the unknown V-periodic displacement correction term satisfying (4.10). The friction condition $(4.78)_5$ relates the leading terms of the microscopic friction stress t_T^0 and sliding velocity $\dot{g}_T^1$, and τ_s^0 is given by (4.74). Thus the boundary value problem (4.78) is a rate problem. Because both the normal component and the tangential component of the velocity $\dot{\mathbf{u}}^1$ affect the contact and friction conditions, equations (4.75) enforce a constraint on $\dot{\mathbf{u}}^1$. Thus, in the macroscopic sticking zone, the solution $\dot{\mathbf{u}}^1$ is uniquely defined by the boundary layer equations (4.78).

The macroscopic friction stress t_T^h has to be specified from the solution of the microscopic problem (4.78). It can easily be checked that the average microscopic friction stress satisfies the following inequality

$$-(\alpha \tau_a - \mu \langle t_N^0 \rangle) \le \langle t_T^0 \rangle \le \alpha \tau_a - \mu \langle t_N^0 \rangle. \tag{4.79}$$

Thus, in view of (4.75), the macroscopic friction condition (4.61) with τ_s^h specified by (4.64), is verified in the case of macroscopic sticking, $\dot{g}_T^h = 0$.

4.6 Conclusions

In this chapter, a general methodology for the analysis of boundary layers induced in linear elastic solids by micro-inhomogeneous boundary conditions has been presented. In addition to the case of micro-inhomogeneous traction and displacement boundary conditions, the consequences of enforcing unilateral contact and friction conditions have been studied for a rough body in contact with a smooth and rigid obstacle. The main results and conclusions following from the analysis in the present chapter are summarized below.

1. The displacement in the boundary layer is formed by a macroscopic deformation $\mathbf{u}^0(\mathbf{x})$ with a superimposed correction term $\mathbf{u}^1(\mathbf{x}', \mathbf{y})$ which is V-periodic, and its gradient vanishes far from the boundary. As a result, the microscopic strain is the sum of the macroscopic strain, $\mathbf{e}_{\mathbf{x}}(\mathbf{u}^0)$, and the strain associated with the displacement correction term, $\mathbf{e}_{\mathbf{y}}(\mathbf{u}^1)$, cf. (4.13).
2. The microscopic problem of the boundary layer is formulated for a half-space. However, in view of periodicity the strip V is the actual domain (periodic unit cell) of the microscopic problem.
3. If the macroscopic friction stress depends on the solution of the microscopic problem, as, for instance, in the case of non-zero adhesive friction component, the macroscopic problem and the microscopic one are coupled and must be solved in parallel.

   Otherwise, the macroscopic problem can be solved independently of the microscopic problem. The microscopic problem is then, merely, a post-processing task for which the solution of the macroscopic problem, i.e. the macroscopic strain $\mathbf{e}_{\mathbf{x}}(\mathbf{u}^0)$, constitutes the input.

4. In the case of micro-inhomogeneous surface traction, the displacement correction $\mathbf{u}^1$ in the boundary layer is only defined up to an additive vector.

In the macroscopic contact zone, the unilateral contact condition enforces a constraint on the normal component of $\mathbf{u}^1$. Otherwise, in the macroscopic separation zone, the solution is defined up to an additive vector.

In the case of frictional contact, the macroscopic contact zone is divided into sticking and sliding zones. In the macroscopic sticking zone, the microscopic problem is a rate problem, and the *velocity* $\dot{\mathbf{u}}^1$ is fully constrained. In the macroscopic sliding zone, the microscopic problem is no longer a rate problem, and only the normal component of the *displacement* correction $\mathbf{u}^1$ is then constrained.

Finally, the displacement correction term $\mathbf{u}^1$ is uniquely defined if a micro-inhomogeneous displacement is prescribed.

As discussed in Point 3 above, in some cases the macroscopic problem and the microscopic one are coupled, so that both problems must be solved simultaneously. If the finite element method were used at both scales to solve the corresponding boundary value problems, the resulting micromechanical scheme would resemble that of the two-scale finite element method, see, for instance, Feyel and Chaboche [30] and Kouznetsova et al. [70].

Alternatively, and this is a common approach in contact mechanics, the analysis of a contact boundary layer can be carried out without referring to a specific macroscopic problem. The effective properties of the contact layer can then be obtained for some typical prescribed loading programs, e.g. by prescribing contact tractions or relative motions of contacting surfaces. The corresponding response of the boundary layer can then be used to derive macroscopic contact laws. This approach is used also in this work, and some examples of micromechanical analysis of contact boundary layers are provided in Chap. 6.

If the Coulomb friction law is assumed to hold at the micro-scale ($\tau_a = 0$), then the macroscopic friction law is also of Coulomb type with the macroscopic friction coefficient equal to the microscopic one, cf. (4.64). This result is obtained for an elastic body in contact with a smooth obstacle. Clearly, if these two assumptions were dropped, then the macroscopic friction coefficient would not necessarily be equal to the microscopic one. Moreover, it might depend on the normal contact pressure, cf. the analysis of asperity ploughing in Sect. 6.3.

With the aim of keeping the notation and the exposition as simple as possible, the boundary layer analysis has been carried out in this chapter for the simplified case of an elastic body in two-dimensions. The present micromechanical framework is further developed in the next chapter for a more general class of problems. In particular, the averaging procedure (4.21) is extensively used, and the properties of the corresponding averages are studied.

5

Micromechanics of Boundary Layers

Abstract: A micromechanical framework is developed for the analysis of deformation inhomogeneities within boundary layers. The main idea is to average the inhomogeneous fields along the surface, but to preserve the dependence of the averages on the distance from the surface. Several properties of such averages are derived. Finally, the analysis of the boundary layer induced in an elastic body by a sinusoidal fluctuation of surface traction is provided as an illustrative example for which analytical solution exists.

5.1 Preliminaries

The present chapter is devoted to micromechanical analysis of boundary layers. The focus is on the microscopic problem of the boundary layer, the macroscopic problem is only referred to as a source of the macroscopic strain. Based on the experience gained in Chap. 4, the equations governing the microscopic problem corresponding to a three-dimensional elasto-plastic body subjected to micro-periodic surface traction[1] are anticipated without detailed derivation. A special averaging operation is also introduced as a basic tool of the present micromechanical analysis, and several properties of boundary layer fields are derived. Finally, in order to illustrate the approach, a simple example of the boundary layer induced in an elastic body by a sinusoidal fluctuation of surface traction is analyzed.

Two results of the boundary layer analysis of Chap. 4 are of particular importance. Firstly, the microscopic boundary value problem is analyzed in a simplified geometry, namely in the half-space, and, in view of periodicity, it is sufficient to consider only the strip V as a unit cell. This is very natural and

---

[1] For brevity, the formulations corresponding to the case of prescribed micro-periodic displacement and to the case of contact interaction with rigid obstacle are not discussed in this chapter.

has been intuitively assumed in numerous micromechanical studies of asperity interaction, as already mentioned in Sect. 4.1. The second result, namely the form of the displacement and strain fields within the boundary layer is less obvious. It provides the link between the macro- and the micro-scale, and thus it enables the analysis of the interaction between the phenomena at the two scales.

In the subsequent sections, the boundary layer equations derived in Chap. 4 are rewritten using a simpler and more natural notation, which is also consistent with the notation used throughout the major part of this book. The analysis is carried out in the scaled spatial variable $\mathbf{y} = \mathbf{x}/\varepsilon$, and the boundary layer correction of displacement, corresponding to $\mathbf{u}^1(\mathbf{x}', \mathbf{y})$ in Chap. 4 and denoted here by $\mathbf{w}(\mathbf{y})$, is also a scaled displacement – the physical displacement is obtained by multiplying the boundary layer correction by ε, cf. (4.7). Furthermore, the dependencies on the spatial macro variable $\mathbf{x}$ are not accounted for, and these dependencies are omitted in the present simplified notation. The spatial derivatives are taken directly by differentiating the fields with respect to $\mathbf{y}$, thus the special rule (4.11), applicable in the case of two-scale analysis, is not used. For instance, the infinitesimal strain associated with a scaled displacement field $\mathbf{u}(\mathbf{y})$ is simply the symmetrized gradient with respect to $\mathbf{y}$,

$$\mathbf{e}(\mathbf{u}) = \frac{1}{2} \left[\operatorname{grad} \mathbf{u} + (\operatorname{grad} \mathbf{u})^T \right]. \tag{5.1}$$

Finally, only the leading terms are accounted for, and thus the superscripts indexing the terms of the asymptotic expansions are dropped. The symbols used in the present chapter, and also in Chap. 6, are listed in Table 5.1 together with the equivalent symbols of Chap. 4.

As shown in Chap. 4, the microscopic problem is a boundary value problem of a homogeneous half-space subjected to a periodic loading on the boundary. Consider thus a half-space in the intrinsic coordinate system such that the

Table 5.1. Correspondence between the simplified notation used in the present chapter and the notation of Chap. 4.

Notation of Chap. 4	Simplified notation	Description
$\mathbf{u}^1(\mathbf{x}', \mathbf{y})$	$\mathbf{w}(\mathbf{y})$	displacement correction in the boundary layer
$\mathbf{e}^0(\mathbf{x}', \mathbf{y})$	$\varepsilon(\mathbf{y})$	microscopic strain
$\mathbf{e_x}(\mathbf{u}^0)$	$\mathbf{E}$	macroscopic strain
$\mathbf{e_y}(\mathbf{u}^1)$	$\mathbf{e}(\mathbf{w})$	strain due to displacement correction
$\boldsymbol{\sigma}^0(\mathbf{x}', \mathbf{y})$	$\boldsymbol{\sigma}(\mathbf{y})$	microscopic stress
$\boldsymbol{\sigma}^h(\mathbf{x})$	$\boldsymbol{\Sigma}$	macroscopic stress
$\operatorname{div}_{\mathbf{y}} \boldsymbol{\sigma}^0 = \mathbf{0}$	$\operatorname{div} \boldsymbol{\sigma} = \mathbf{0}$	equilibrium equation
$\mathbf{t}^0, t_N^0, t_T^0$	$\mathbf{t}, t_N, \mathbf{t}_T$	microscopic contact tractions
$\mathbf{t}^h, t_N^h, t_T^h$	$\mathbf{T}, T_N, \mathbf{T}_T$	macroscopic contact tractions

boundary surface is given by $y_3 = 0$ and the y_3-axis is directed into the half-space. Further, denote by S the unit surface element within the (y_1, y_2)-plane and by $\boldsymbol{\eta} = (y_1, y_2)$ the position within the (y_1, y_2)-plane. A function $f(\boldsymbol{\eta})$ will be called S-*periodic* if

$$f(\boldsymbol{\eta}^+) = f(\boldsymbol{\eta}^-), \qquad \boldsymbol{\eta}^+ \in \partial S^+,\ \boldsymbol{\eta}^- \in \partial S^-, \tag{5.2}$$

where, the boundary ∂S is divided into two parts, ∂S^+ and ∂S^-, and $\boldsymbol{\eta}^+ \in \partial S^+$ and $\boldsymbol{\eta}^- \in \partial S^-$ are two associated points with opposite outward normals, cf. Fig. 5.1. The unit surface element S defines the unit element of the half-space, namely the half-infinite strip V, cf. Fig. 5.1,

$$V = \{\mathbf{y} = (\boldsymbol{\eta}, y_3) \colon y_3 \geq 0,\ \boldsymbol{\eta} \in S\}. \tag{5.3}$$

A function $f(\mathbf{y})$ defined on the strip V will be called V-*periodic* if

$$f(\mathbf{y}^+) = f(\mathbf{y}^-), \qquad \mathbf{y}^+ = (\boldsymbol{\eta}^+, y_3),\ \mathbf{y}^- = (\boldsymbol{\eta}^-, y_3). \tag{5.4}$$

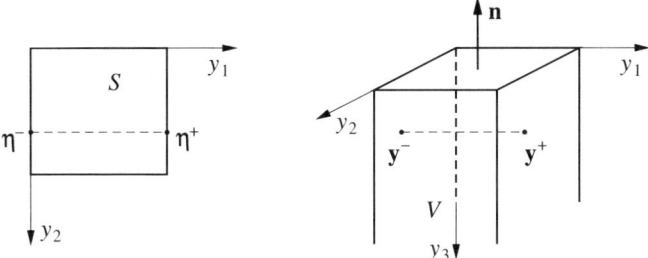

Fig. 5.1. Unit surface element S and strip V in three-dimensional case.

5.2 Averaging and Micromechanical Relations

5.2.1 Averaging Operation

Let us define an averaging operation over the unit surface element S, which, for a field $\varphi(\mathbf{y})$, $\mathbf{y} \in V$, introduces the average $\bar{\varphi}(y_3)$, corresponding to a fixed distance y_3 from the boundary $y_3 = 0$, and the fluctuation $\tilde{\varphi}(\mathbf{y})$, namely

$$\varphi(\mathbf{y}) = \bar{\varphi}(y_3) + \tilde{\varphi}(\mathbf{y}), \qquad \bar{\varphi} = \langle \varphi \rangle, \qquad \langle \tilde{\varphi} \rangle = 0, \tag{5.5}$$

where

$$\langle \varphi \rangle(y_3) \equiv \frac{1}{|S|} \int_S \varphi(\boldsymbol{\eta}, y_3)\, \mathrm{d}S, \qquad |S| = \int_S \mathrm{d}S, \tag{5.6}$$

see also Sect. 4.2.5.

Consider also a slice $V_h(y_3')$ of thickness $h > 0$, positioned at $y_3 = y_3'$, so that $y_3' \leq y_3 \leq y_3' + h$, and introduce a volume average within $V_h(y_3')$, defined by

$$\{\varphi\}_h(y_3') \equiv \frac{1}{h|S|} \int_{V_h(y_3')} \varphi(\mathbf{y}) \, dV = \frac{1}{h|S|} \int_{y_3'}^{y_3'+h} \int_S \varphi(\boldsymbol{\eta}, y_3) \, dS \, dy_3. \quad (5.7)$$

Hence in the limit,

$$\langle\varphi\rangle(y_3) = \lim_{h\to 0} \{\varphi\}_h(y_3). \quad (5.8)$$

5.2.2 Kinematically Admissible Displacement Field

As shown in Sect. 4.2, the strain in the boundary layer (the leading term of the asymptotic expansion) is a sum of the macroscopic strain and the strain associated with the displacement correction, cf. (4.12)–(4.13). In the present notation this is written as

$$\boxed{\varepsilon(\mathbf{y}) = \mathbf{E} + \mathbf{e}(\mathbf{w}(\mathbf{y})),} \quad (5.9)$$

where $\mathbf{E}$ is the macroscopic strain, $\mathbf{w}(\mathbf{y})$ is the V-periodic displacement correction, and $\mathbf{e}(\mathbf{w})$ is the strain component associated with the displacement correction $\mathbf{w}(\mathbf{y})$, cf. (5.1).

Referring to the boundary layer analysis of Chap. 4, the displacement field $\mathbf{w}(\mathbf{y})$ is a boundary layer correction superimposed on the micro-homogeneous macroscopic displacement field. However, the macroscopic displacement affects the microscopic problem only through the macroscopic strain $\mathbf{E}$. Accordingly, for the present purpose, the total displacement $\mathbf{u}(\mathbf{y})$ within the boundary layer can be formed by considering only the linear term of the macroscopic displacement, namely

$$\boxed{\mathbf{u}(\mathbf{y}) = \mathbf{u}_0 + \mathbf{H}\mathbf{y} + \mathbf{w}(\mathbf{y}),} \quad (5.10)$$

where

$$\mathbf{H} = \mathbf{E} + \boldsymbol{\Omega}, \qquad \boldsymbol{\Omega} = -\boldsymbol{\Omega}^T. \quad (5.11)$$

Here $\mathbf{u}_0$ is an arbitrary vector, and $\boldsymbol{\Omega}$ is an arbitrary antisymmetric tensor associated with rigid body rotation, see also Remark 5.1. The analogous term related to $\mathbf{w}(\mathbf{y})$ is ruled out by V-periodicity of $\mathbf{w}(\mathbf{y})$. Clearly, the strain $\varepsilon(\mathbf{y})$ in the form specified by (5.9) derives from the displacement field (5.10), so that

$$\varepsilon(\mathbf{y}) = \mathbf{e}(\mathbf{u}). \quad (5.12)$$

The boundary layer correction $\mathbf{w}(\mathbf{y})$, $\mathbf{y} \in V$, will be called a *kinematically admissible correction*, and the displacement field $\mathbf{u}(\mathbf{y})$ of the form (5.10) will be called a *kinematically admissible displacement*, if

i. $\mathbf{w}(\mathbf{y})$ is V-periodic,

ii. and its gradient vanishes far from the surface, cf. (4.10),

$$\operatorname{grad} \mathbf{w}(\mathbf{y}) \xrightarrow{y_3 \to +\infty} \mathbf{0}. \tag{5.13}$$

It follows from (5.9) and (5.13) that the strain $\varepsilon(\mathbf{y}) = \mathbf{e}(\mathbf{u})$ derived from a kinematically admissible displacement $\mathbf{u}(\mathbf{y})$ satisfies

$$\varepsilon(\mathbf{y}) \xrightarrow{y_3 \to +\infty} \mathbf{E}, \tag{5.14}$$

i.e., far from the boundary, the strain $\varepsilon(\mathbf{y})$ in the boundary layer *matches* the macroscopic strain $\mathbf{E}$.

Consider now the average strain $\bar{\varepsilon}(y_3)$ in the boundary layer, which is obtained by averaging $\varepsilon(\mathbf{y})$, cf. (5.9), namely

$$\bar{\varepsilon}(y_3) = \langle \varepsilon \rangle (y_3) = \mathbf{E} + \mathbf{e}(\bar{\mathbf{w}}(y_3)). \tag{5.15}$$

Since $\bar{\mathbf{w}}$ depends only on y_3, its derivatives with respect to y_1 and y_2 vanish. Accordingly, $\mathbf{e}(\bar{\mathbf{w}})$ can be expressed as a symmetrized diadic product,

$$\mathbf{e}(\bar{\mathbf{w}}) = \tfrac{1}{2}(\bar{\mathbf{w}}_{,3} \otimes \mathbf{n} + \mathbf{n} \otimes \bar{\mathbf{w}}_{,3}), \tag{5.16}$$

where $\bar{\mathbf{w}}_{,3} = d\bar{\mathbf{w}}/dy_3$, and thus the *interior* part of $\mathbf{e}(\bar{\mathbf{w}})$ vanishes, cf. Sect. 2.2. This can be written in the form of the following compatibility condition,

$$\boxed{\bar{\varepsilon}_{\mathrm{P}}(y_3) = \mathbf{E}_{\mathrm{P}} = \mathrm{const},} \tag{5.17}$$

which is a fundamental property of the kinematics of boundary layers.

Since $\mathbf{w} = \bar{\mathbf{w}} + \tilde{\mathbf{w}}$, the strain fluctuation $\tilde{\varepsilon}(\mathbf{y})$ is found to be the strain derived from $\tilde{\mathbf{w}}(\mathbf{y})$, i.e. from the fluctuation of displacement correction $\mathbf{w}(\mathbf{y})$,

$$\boxed{\tilde{\varepsilon}(\mathbf{y}) = \mathbf{e}(\tilde{\mathbf{w}}(\mathbf{y})).} \tag{5.18}$$

Remark 5.1. The choice of $\boldsymbol{\Omega}$ in (5.11) is arbitrary. However, in some situations, e.g. in the case of contact boundary layers studied in Chap. 6, it is convenient to assume $\boldsymbol{\Omega}$ in the following form

$$\boldsymbol{\Omega} = (\mathbf{En}) \otimes \mathbf{n} - \mathbf{n} \otimes (\mathbf{En}), \tag{5.19}$$

so that in terms of components we have

$$H_{ij} = \begin{pmatrix} E_{11} & E_{12} & 2E_{13} \\ E_{12} & E_{22} & 2E_{23} \\ 0 & 0 & E_{33} \end{pmatrix}. \tag{5.20}$$

Then, the part of the y_3-displacement that is related to the macroscopic strain $\mathbf{E}$ does not depend on y_1 and y_2. The corresponding deformation of the strip V is schematically shown in Fig. 5.2.

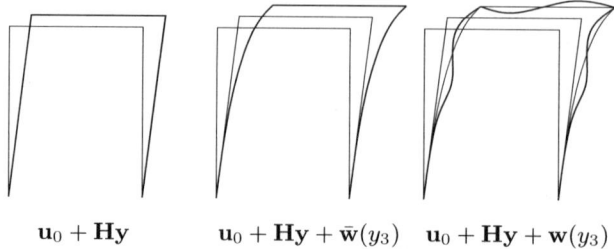

$$\mathbf{u}_0 + \mathbf{Hy} \qquad \mathbf{u}_0 + \mathbf{Hy} + \bar{\mathbf{w}}(y_3) \qquad \mathbf{u}_0 + \mathbf{Hy} + \mathbf{w}(y_3)$$

Fig. 5.2. Displacement field within the boundary layer (schematic).

5.2.3 Statically Admissible Stress Field

A stress field $\boldsymbol{\sigma}(\mathbf{y})$ will be called *statically admissible* if

i. $\boldsymbol{\sigma}(\mathbf{y})$ satisfies the equilibrium equation within the strip V,

$$\operatorname{div} \boldsymbol{\sigma} = \mathbf{0}; \tag{5.21}$$

ii. the surface tractions $\mathbf{t}^{\pm} = \boldsymbol{\sigma}^{\pm} \mathbf{n}^{\pm}$ are anti-periodic,

$$\boldsymbol{\sigma}(\mathbf{y}^+)\mathbf{n}^+ = -\boldsymbol{\sigma}(\mathbf{y}^-)\mathbf{n}^-, \qquad \mathbf{y}^+ = (\boldsymbol{\eta}^+, y_3), \ \mathbf{y}^- = (\boldsymbol{\eta}^-, y_3), \tag{5.22}$$

where $\mathbf{n}^+$ and $\mathbf{n}^- = -\mathbf{n}^+$ are outer normals to ∂V at, respectively, $\mathbf{y}^+$ and $\mathbf{y}^-$, and $\boldsymbol{\eta}^+ \in \partial S^+$ and $\boldsymbol{\eta}^- \in \partial S^-$;

iii. $\boldsymbol{\sigma}(\mathbf{y})$ satisfies the boundary condition at $y_3 = 0$,

$$\boldsymbol{\sigma}(\mathbf{y})\mathbf{n} = \mathbf{t}^{\mathrm{b}}(\boldsymbol{\eta}) \quad \text{for } y_3 = 0, \ \boldsymbol{\eta} \in S, \tag{5.23}$$

where $\mathbf{t}^{\mathrm{b}}(\boldsymbol{\eta})$ is a prescribed S-periodic surface traction.

An important property of statically admissible stress fields has been derived in Sect. 4.2, cf. (4.24). In the present notation, equation (4.24) reads

$$\boxed{\bar{\boldsymbol{\sigma}}_{\mathrm{A}}(y_3) = \boldsymbol{\Sigma}_{\mathrm{A}} = \text{const},} \tag{5.24}$$

and expresses the overall equilibrium of the strip V or any slice $V_h(y_3)$. Equation (5.24) can be equivalently written in the following form,

$$\bar{\boldsymbol{\sigma}}(y_3)\mathbf{n} = \boldsymbol{\Sigma}\mathbf{n} = \bar{\mathbf{t}}^{\mathrm{b}}, \tag{5.25}$$

as a counterpart to (4.23), where $\bar{\mathbf{t}}^{\mathrm{b}} = \langle \mathbf{t}^{\mathrm{b}} \rangle$ is the average (macroscopic) surface traction.

By averaging the equilibrium equation (5.21), and by decomposing a statically admissible stress field $\boldsymbol{\sigma}(\mathbf{y})$ into its average $\bar{\boldsymbol{\sigma}}(y_3)$ and fluctuation $\tilde{\boldsymbol{\sigma}}(\mathbf{y})$, we observe that both fields are *statically admissible*. Indeed, we have

$$\operatorname{div} \bar{\boldsymbol{\sigma}} = \mathbf{0}, \qquad \operatorname{div} \tilde{\boldsymbol{\sigma}} = \mathbf{0}, \tag{5.26}$$

the periodicity condition (5.22) is trivially satisfied by both $\bar{\boldsymbol{\sigma}}$ and $\tilde{\boldsymbol{\sigma}}$, and the boundary condition at $y_3 = 0$ is satisfied in the following sense,

$$\bar{\boldsymbol{\sigma}}(y_3)\mathbf{n} = \bar{\mathbf{t}}^{\mathrm{b}}, \quad \tilde{\boldsymbol{\sigma}}(\mathbf{y})\mathbf{n} = \tilde{\mathbf{t}}^{\mathrm{b}}(\boldsymbol{\eta}) \quad \text{for } y_3 = 0, \ \boldsymbol{\eta} \in S, \tag{5.27}$$

where $\mathbf{t}^{\mathrm{b}}(\boldsymbol{\eta}) = \bar{\mathbf{t}}^{\mathrm{b}} + \tilde{\mathbf{t}}^{\mathrm{b}}(\boldsymbol{\eta})$.

5.2.4 Compatibility Conditions for Averages

Let $\Delta\boldsymbol{\sigma}$ and $\Delta\boldsymbol{\varepsilon}$ denote the change of, respectively, stress and strain in the boundary layer with respect to the corresponding macroscopic values, viz.

$$\Delta\boldsymbol{\varepsilon} = \boldsymbol{\varepsilon} - \mathbf{E}, \quad \Delta\boldsymbol{\sigma} = \boldsymbol{\sigma} - \boldsymbol{\Sigma}. \tag{5.28}$$

The *compatibility conditions for averages*, specified by equations (5.17) and (5.24), which hold for a kinematically admissible displacement field and for a statically admissible stress field, respectively, can now be written in the form

$$\boxed{\Delta\bar{\boldsymbol{\varepsilon}}_{\mathrm{P}} = \mathbf{0}, \quad \Delta\bar{\boldsymbol{\sigma}}_{\mathrm{A}} = \mathbf{0},} \tag{5.29}$$

or in an equivalent form

$$\Delta\bar{\boldsymbol{\varepsilon}} = \Delta\bar{\boldsymbol{\varepsilon}}_{\mathrm{A}}, \quad \Delta\bar{\boldsymbol{\sigma}} = \Delta\bar{\boldsymbol{\sigma}}_{\mathrm{P}}, \tag{5.30}$$

where, in view of (5.16),

$$\Delta\bar{\boldsymbol{\varepsilon}} = \mathbf{e}(\bar{\mathbf{w}}) = \tfrac{1}{2}(\bar{\mathbf{w}}_{,3} \otimes \mathbf{n} + \mathbf{n} \otimes \bar{\mathbf{w}}_{,3}). \tag{5.31}$$

Furthermore, the above compatibility conditions imply that $\Delta\bar{\boldsymbol{\varepsilon}}$ and $\Delta\bar{\boldsymbol{\sigma}}$ are orthogonal, cf. the property (2.5) of the interior–exterior decomposition, so that

$$\boxed{\Delta\bar{\boldsymbol{\sigma}} \cdot \Delta\bar{\boldsymbol{\varepsilon}} = 0.} \tag{5.32}$$

We note that the compatibility conditions (5.29) and the orthogonality condition (5.32) formulated for the average stress and strain in the boundary layer are formally identical to the respective conditions that hold locally for the jumps of stress and strain at a bonded interface, cf. (2.30) and (2.31).

5.2.5 Average Work

Let us now consider the average work of a statically admissible stress $\boldsymbol{\sigma}(\mathbf{y})$ on strain $\boldsymbol{\varepsilon}(\mathbf{y})$ derived from a kinematically admissible displacement $\mathbf{u}(\mathbf{y})$. In view of $\langle \bar{\boldsymbol{\sigma}} \cdot \tilde{\boldsymbol{\varepsilon}} \rangle = \bar{\boldsymbol{\sigma}} \cdot \langle \tilde{\boldsymbol{\varepsilon}} \rangle = 0$ and $\langle \tilde{\boldsymbol{\sigma}} \cdot \bar{\boldsymbol{\varepsilon}} \rangle = \langle \tilde{\boldsymbol{\sigma}} \rangle \cdot \bar{\boldsymbol{\varepsilon}} = 0$, we have

$$\langle \boldsymbol{\sigma} \cdot \boldsymbol{\varepsilon} \rangle = \bar{\boldsymbol{\sigma}} \cdot \bar{\boldsymbol{\varepsilon}} + \langle \tilde{\boldsymbol{\sigma}} \cdot \tilde{\boldsymbol{\varepsilon}} \rangle. \tag{5.33}$$

In order to evaluate the term $\langle \tilde{\boldsymbol{\sigma}} \cdot \tilde{\boldsymbol{\varepsilon}} \rangle$ in (5.33), we consider the weak form of the equilibrium equation formulated for stress fluctuation $\tilde{\boldsymbol{\sigma}}(\mathbf{y})$. It has been shown that, if a stress field $\boldsymbol{\sigma}(\mathbf{y})$ is statically admissible, so is its fluctuation $\tilde{\boldsymbol{\sigma}}(\mathbf{y})$, cf. $(5.26)_2$. Adopting $\tilde{\mathbf{w}}$ as a test function, the weak form of equilibrium of a slice $V_h(y_3)$ is

$$\int_{V_h(y_3)} \tilde{\boldsymbol{\sigma}} \cdot \tilde{\boldsymbol{\varepsilon}} \, dV = \int_{\partial V_h(y_3)} \tilde{\mathbf{t}} \cdot \tilde{\mathbf{w}} \, dS, \qquad (5.34)$$

where $\tilde{\boldsymbol{\varepsilon}} = \mathbf{e}(\tilde{\mathbf{w}})$, $\tilde{\mathbf{t}} = \tilde{\boldsymbol{\sigma}}\mathbf{n}'$, and $\mathbf{n}'(\mathbf{y})$ is the outer normal to $V_h(y_3)$. Dividing (5.34) by the volume of $V_h(y_3)$, we obtain

$$\{\tilde{\boldsymbol{\sigma}} \cdot \tilde{\boldsymbol{\varepsilon}}\}_h(y_3) = \frac{1}{h|S|} \int_{\partial V_h(y_3)} \tilde{\mathbf{t}} \cdot \tilde{\mathbf{w}} \, dS. \qquad (5.35)$$

We shall now consider the limit of (5.35) as $h \to 0$. In view of (5.8) the left-hand side of (5.35) becomes

$$\lim_{h \to 0} \{\tilde{\boldsymbol{\sigma}} \cdot \tilde{\boldsymbol{\varepsilon}}\}_h(y_3) = \langle \tilde{\boldsymbol{\sigma}} \cdot \tilde{\boldsymbol{\varepsilon}} \rangle(y_3). \qquad (5.36)$$

The integral on the right-hand side of (5.35) can be split into four parts,

$$\frac{1}{h|S|} \int_{\partial V_h(y_3)} \tilde{\mathbf{t}} \cdot \tilde{\mathbf{w}} \, dS = \frac{1}{h} \langle \tilde{\mathbf{w}} \cdot \tilde{\boldsymbol{\sigma}}\mathbf{n} \rangle(y_3)$$
$$- \frac{1}{h} \langle \tilde{\mathbf{w}} \cdot \tilde{\boldsymbol{\sigma}}\mathbf{n} \rangle(y_3 + h)$$
$$+ \int_{y_3}^{y_3+h} \int_{\partial S^+} \tilde{\mathbf{t}} \cdot \tilde{\mathbf{w}} \, ds \, dy_3'$$
$$+ \int_{y_3}^{y_3+h} \int_{\partial S^-} \tilde{\mathbf{t}} \cdot \tilde{\mathbf{w}} \, ds \, dy_3'. \qquad (5.37)$$

Here, the first and the second term correspond to the parts of the boundary $\partial V_h(y_3)$ parallel to the surface, so that the normal vector is $\mathbf{n}' = \pm\mathbf{n}$. The remaining terms, corresponding to lateral faces of $\partial V_h(y_3)$, cancel each other due to periodicity of $\tilde{\mathbf{w}}$ and anti-periodicity of $\tilde{\mathbf{t}}$. Thus, in the limit of $h \to 0$, we have

$$\lim_{h \to 0} \frac{1}{h|S|} \int_{\partial V_h(y_3)} \tilde{\mathbf{t}} \cdot \tilde{\mathbf{w}} \, dS = -\frac{d}{dy_3} \langle \tilde{\mathbf{w}} \cdot \tilde{\boldsymbol{\sigma}}\mathbf{n} \rangle, \qquad (5.38)$$

and equations (5.34)–(5.38) yield

$$\langle \tilde{\boldsymbol{\sigma}} \cdot \tilde{\boldsymbol{\varepsilon}} \rangle = -\frac{d}{dy_3} \langle \tilde{\mathbf{w}} \cdot \tilde{\boldsymbol{\sigma}}\mathbf{n} \rangle. \qquad (5.39)$$

Finally, by combining (5.33) and (5.39), the average work is obtained in the form

$$\boxed{\langle \boldsymbol{\sigma} \cdot \boldsymbol{\varepsilon} \rangle = \bar{\boldsymbol{\sigma}} \cdot \bar{\boldsymbol{\varepsilon}} - \frac{d}{dy_3} \langle \tilde{\mathbf{w}} \cdot \tilde{\boldsymbol{\sigma}}\mathbf{n} \rangle.} \qquad (5.40)$$

Note that, unlike in the classical micromechanics, cf. the Hill's lemma (2.14), the average work *is not equal* to the work of averages.

5.3 Boundary Value Problem

Using the notion of the kinematically admissible displacement and statically admissible stress fields, the boundary value problem can be stated by simply requiring that admissible stress and strain fields satisfy the constitutive equation.

5.3.1 Linear Elasticity

In the case of a linear elastic body, the microscopic problem is thus to find a kinematically admissible correction $\mathbf{w}(\mathbf{y})$, $\mathbf{y} \in V$, such that the stress field $\boldsymbol{\sigma}(\mathbf{y})$, given by

$$\boldsymbol{\sigma}(\mathbf{y}) = \mathbf{L}\varepsilon(\mathbf{y}), \qquad \varepsilon(\mathbf{y}) = \mathbf{E} + \mathbf{e}(\mathbf{w}), \tag{5.41}$$

is statically admissible, where the macroscopic strain $\mathbf{E}$ and the S-periodic surface traction $\mathbf{t}^b(\boldsymbol{\eta})$ are given. The material is assumed to be elastically homogeneous, i.e. the elastic stiffness tensor $\mathbf{L}$ is constant within V.

Importantly, the prescribed macroscopic strain $\mathbf{E}$ and the surface traction $\mathbf{t}^b(\boldsymbol{\eta})$ are not completely independent. This is because the average traction $\bar{\mathbf{t}}^b = \langle \mathbf{t}^b \rangle$ and the macroscopic stress $\boldsymbol{\Sigma} = \mathbf{L}\mathbf{E}$ must satisfy the compatibility condition (5.25) expressing the overall equilibrium of strip V. However, this is guaranteed if the macroscopic strain $\mathbf{E}$ follows from the solution of the macroscopic problem.

As the elastic stiffness tensor $\mathbf{L}$ is constant, the average stress $\bar{\boldsymbol{\sigma}}(y_3)$ and the average strain $\bar{\varepsilon}(y_3)$ also satisfy the constitutive equation

$$\bar{\boldsymbol{\sigma}}(y_3) = \mathbf{L}\bar{\varepsilon}(y_3), \tag{5.42}$$

which is obtained by averaging the constitutive equation (5.41). In view of the compatibility conditions (5.29), the average stress and the average strain in the boundary layer are equal to the respective macroscopic quantities,

$$\Delta\bar{\varepsilon}(y_3) = \mathbf{0}, \qquad \Delta\bar{\boldsymbol{\sigma}}(y_3) = \mathbf{0}, \tag{5.43}$$

which follows, for instance, from the mixed form of constitutive equation (3.3) with $\varepsilon^t = \mathbf{0}$. Furthermore, in view of $(5.43)_1$ and (5.31), we have

$$\bar{\mathbf{w}}(y_3) = \text{const}, \tag{5.44}$$

i.e. the displacement correction in the elastic boundary layer is, essentially, a pure fluctuation, $\mathbf{w}(\mathbf{y}) = \tilde{\mathbf{w}}(\mathbf{y}) + \text{const}$.

5.3.2 Elasto-Plasticity

Linear elasticity has been assumed throughout Chap. 4 and the preceding sections of the present chapter. Rigorous extension of the above analysis to the elasto-plastic case is not presented here. The notions of admissible displacements and stresses are thus adopted as defined in Sect. 5.2, and the analysis follows by specifying the constitutive equations of elasto-plasticity.

Consider thus an elasto-plastic boundary layer. The strain ε is then decomposed into elastic ε^e and plastic ε^p parts and the constitutive equation,

$$\boldsymbol{\sigma}(\mathbf{y}) = \mathbf{L}\varepsilon^e(\mathbf{y}), \qquad \varepsilon^e(\mathbf{y}) = \varepsilon(\mathbf{y}) - \varepsilon^p(\mathbf{y}), \tag{5.45}$$

is accompanied by the yield condition

$$F(\boldsymbol{\sigma}) = \sqrt{\frac{3}{2}\,\mathbf{s}\cdot\mathbf{s}} - \sigma_y(\varepsilon^p) \leq 0, \qquad \mathbf{s} = \boldsymbol{\sigma} - \frac{1}{3}\mathbf{I}\,\mathrm{tr}\,\boldsymbol{\sigma}, \tag{5.46}$$

and the associated flow rule

$$\dot{\varepsilon}^p = \gamma\frac{\partial F}{\partial\boldsymbol{\sigma}}, \qquad \gamma \geq 0, \qquad \gamma F = 0. \tag{5.47}$$

Here $\mathbf{s}(\mathbf{y})$ is the stress deviator, and σ_y is the uniaxial yield stress which is a given function of the (local) equivalent plastic strain $\varepsilon^p(\mathbf{y})$, with the evolution law $\dot{\varepsilon}^p = \gamma = (\frac{2}{3}\dot{\varepsilon}^p\cdot\dot{\varepsilon}^p)^{1/2}$. Clearly, the macroscopic quantities also satisfy the above constitutive equations (5.45)–(5.47), expressed in terms of the macroscopic stress $\boldsymbol{\Sigma}$ and the macroscopic strain $\mathbf{E} = \mathbf{E}^e + \mathbf{E}^p$, where $\mathbf{E}^e$ and $\mathbf{E}^p$ are, respectively, the elastic and plastic parts of the macroscopic strain $\mathbf{E}$.

The microscopic problem is now a problem of evolution in time. For a S-periodic surface traction $\mathbf{t}^b(\boldsymbol{\eta}, t)$ and for a macroscopic strain $\mathbf{E}(t)$, both being now given[2] functions of time t, the problem is to find a kinematically admissible correction $\mathbf{w}(\mathbf{y}, t)$, $\mathbf{y} \in V$, $t \in (0, T)$, such that the stress field $\boldsymbol{\sigma}(\mathbf{y}, t)$, satisfying constitutive equations (5.45)–(5.47), is statically admissible.

The properties of elasto-plastic boundary layers can now be analyzed. In view of homogeneity of the elastic moduli tensor $\mathbf{L}$, averaging of the constitutive equation (5.45) gives a simple result, namely

$$\bar{\sigma}(y_3) = \mathbf{L}[\bar{\varepsilon}(y_3) - \bar{\varepsilon}^p(y_3)], \tag{5.48}$$

where $\bar{\varepsilon}^p = \langle\bar{\varepsilon}^p\rangle$. Now, using the compatibility conditions (5.29) and the averaged constitutive equation (5.48), the average stress $\bar{\sigma}(y_3)$ and the average strain $\bar{\varepsilon}(y_3)$ can be uniquely determined in terms of the average plastic strain $\bar{\varepsilon}^p(y_3)$ and macroscopic quantities $\boldsymbol{\Sigma}$, $\mathbf{E}$, and $\mathbf{E}^p$. Indeed, identifying the compatibility conditions (5.29) and the averaged constitutive equation (5.48) with the respective local counterparts, cf. (2.32) and (2.30), the following relationships are obtained as a special case of the interfacial relationships (2.33),

$$\boxed{\Delta\bar{\varepsilon}(y_3) = \mathbf{P}^0\mathbf{L}\Delta\bar{\varepsilon}^p(y_3), \qquad \Delta\bar{\sigma}(y_3) = -\mathbf{S}^0\Delta\bar{\varepsilon}^p(y_3),} \tag{5.49}$$

---

[2] As discussed in Sect. 5.3.1, the average surface traction $\bar{\mathbf{t}}^b$ and the macroscopic strain $\mathbf{E}$ are not independent.

where $\Delta\bar{\varepsilon}^{\mathrm{p}}(y_3) = \bar{\varepsilon}^{\mathrm{p}}(y_3) - \mathbf{E}^{\mathrm{p}}$, and operators $\mathbf{P}^0$ and $\mathbf{S}^0$ are expressed in terms of the elastic moduli tensor $\mathbf{L}$ and normal vector $\mathbf{n}$. It can easily be verified that $\Delta\bar{\varepsilon}$ and $\Delta\bar{\sigma}$ predicted by (5.49) do not violate the compatibility conditions for averages (5.29) due to the special form of operators $\mathbf{P}^0$ and $\mathbf{S}^0$, see equation (A.10).

Consider now the averaged yield condition (5.46). After decomposing the stress deviator $\mathbf{s}(\mathbf{y})$ into its average $\bar{\mathbf{s}}(y_3)$ and fluctuation $\tilde{\mathbf{s}}(\mathbf{y})$, we have $\langle \mathbf{s}\cdot\mathbf{s}\rangle = \bar{\mathbf{s}}\cdot\bar{\mathbf{s}} + \langle \tilde{\mathbf{s}}\cdot\tilde{\mathbf{s}}\rangle$, and from (5.46) it follows that

$$\frac{3}{2}\bar{\mathbf{s}}\cdot\bar{\mathbf{s}} \leq \langle\sigma_{\mathrm{y}}^2\rangle - \frac{3}{2}\langle\tilde{\mathbf{s}}\cdot\tilde{\mathbf{s}}\rangle. \tag{5.50}$$

Inequality (5.50) can be written in the form of an effective yield condition for the average stress $\bar{\sigma}(y_3)$, namely

$$F^{\mathrm{eff}}(\bar{\sigma}) = \sqrt{\frac{3}{2}\bar{\mathbf{s}}\cdot\bar{\mathbf{s}}} - \sigma_{\mathrm{y}}^{\mathrm{eff}} \leq 0, \qquad \bar{\mathbf{s}} = \bar{\sigma} - \frac{1}{3}\mathbf{I}\operatorname{tr}\bar{\sigma}, \tag{5.51}$$

where the effective yield stress $\sigma_{\mathrm{y}}^{\mathrm{eff}}(y_3)$ is defined by

$$\sigma_{\mathrm{y}}^{\mathrm{eff}} = \sqrt{\langle\sigma_{\mathrm{y}}^2\rangle - \frac{3}{2}\langle\tilde{\mathbf{s}}\cdot\tilde{\mathbf{s}}\rangle}. \tag{5.52}$$

A simple bound on the effective yield stress $\sigma_{\mathrm{y}}^{\mathrm{eff}}(y_3)$ follows from (5.52), namely

$$\boxed{\sigma_{\mathrm{y}}^{\mathrm{eff}} < \sqrt{\langle\sigma_{\mathrm{y}}^2\rangle} \qquad \text{if } \tilde{\mathbf{s}} \neq \mathbf{0},} \tag{5.53}$$

and in the case of a perfectly plastic material we have

$$\boxed{\sigma_{\mathrm{y}}^{\mathrm{eff}} < \sigma_{\mathrm{y}} \qquad \text{if } \tilde{\mathbf{s}} \neq \mathbf{0} \text{ and } \sigma_{\mathrm{y}} = \text{const.}} \tag{5.54}$$

It follows from equations (5.51)–(5.54) that, in terms of the average stress $\bar{\sigma}$, the boundary layer is *weakened* due to inhomogeneity of deformation. The weakening of contact boundary layers plays a key role in the phenomenological modelling presented in Sect. 3.3.

5.4 Example: Elastic Boundary Layer Induced by Sinusoidal Traction

As an application of the micromechanical framework developed above, in this section, a boundary layer induced by a periodic surface traction in an elastic body is considered. Specifically, a sinusoidal fluctuation of the normal traction is studied as a simple and illustrative example for which an analytical solution exists.

Consider first an isotropic elastic half-space subjected to sinusoidal normal surface traction

$$p(y_1) = p^* \cos(2\pi y_1/L) \qquad (5.55)$$

where p^* is the amplitude, and L is the wavelength. In the plane strain conditions ($\varepsilon_{22} = \varepsilon_{12} = \varepsilon_{23} = 0$), the stresses within the half-space can be derived from the stress function provided by Johnson [56],

$$\phi(y_1, y_3) = \frac{p^*}{\alpha^2}(1 + \alpha y_3)\,e^{-\alpha y_3}\cos\alpha y_1, \qquad \alpha = 2\pi/L, \qquad (5.56)$$

according to

$$\sigma_{11} = \frac{\partial^2\phi}{\partial y_3^2}, \qquad \sigma_{33} = \frac{\partial^2\phi}{\partial y_1^2}, \qquad \sigma_{13} = -\frac{\partial^2\phi}{\partial y_1 \partial y_3}, \qquad (5.57)$$

and the displacements at the surface are given by

$$
\begin{aligned}
u_1(y_1, 0) &= -\frac{2p^*}{\alpha E}(1 - 2\nu)(1 + \nu)\sin(\alpha y_1) + C_1, \\
u_3(y_1, 0) &= \frac{2p^*}{\alpha E}(1 - \nu^2)\cos(\alpha y_1) + C_2.
\end{aligned}
\qquad (5.58)
$$

where C_1 and C_2 are integration constants.

Consider now an elastic boundary layer subjected to surface tractions specified by

$$t_1^{\mathrm{b}} = \bar{q}, \qquad t_3^{\mathrm{b}}(y_1) = \bar{p} + p^* \cos(2\pi y_1/L), \qquad (5.59)$$

where t_i^{b} are the components of the traction vector $\mathbf{t}^{\mathrm{b}}$, $\bar{p}$ and $\bar{q}$ are the average normal and tangential tractions, and p^* is now the amplitude of the sinusoidal fluctuation of the normal traction.

Let $\boldsymbol{\Sigma}$ be the macroscopic stress. From the compatibility condition (5.25), we have $\Sigma_{13} = \bar{q}$ and $\Sigma_{33} = -\bar{p}$, while the component Σ_{11} cannot be determined from surface data, and it results from the solution of the macroscopic problem. Using equations (5.56)–(5.57), the stresses within the boundary layer are found to be

$$
\begin{aligned}
\sigma_{11} &= \Sigma_{11} - p^*(1 - \alpha y_3)\,e^{-\alpha y_3}\cos\alpha y_1, \\
\sigma_{33} &= \Sigma_{33} - p^*(1 + \alpha y_3)\,e^{-\alpha y_3}\cos\alpha y_1, \\
\sigma_{13} &= \Sigma_{13} - p^*\alpha y_3\,e^{-\alpha y_3}\sin\alpha y_1,
\end{aligned}
\qquad (5.60)
$$

and the strains are obtained from the Hooke's law,

$$
\begin{aligned}
\varepsilon_{11} &= \mathrm{E}_{11} - (p^*/E)(1 + \nu)(1 - 2\nu - \alpha y_3)\,e^{-\alpha y_3}\cos\alpha y_1, \\
\varepsilon_{33} &= \mathrm{E}_{33} - (p^*/E)(1 + \nu)(1 - 2\nu + \alpha y_3)\,e^{-\alpha y_3}\cos\alpha y_1, \\
\varepsilon_{13} &= \mathrm{E}_{13} - (p^*/E)(1 + \nu)\,\alpha y_3\,e^{-\alpha y_3}\sin\alpha y_1,
\end{aligned}
\qquad (5.61)
$$

where E_{ij} are the components of the macroscopic strain $\mathbf{E}$.

It follows from equations (5.60) and (5.61) that, as expected, the stresses and the strains are periodic with respect to y_1. Secondly, it is easily verified that the average stresses and strains are equal to the macroscopic ones, $\langle \boldsymbol{\sigma} \rangle = \boldsymbol{\Sigma}$ and $\langle \boldsymbol{\varepsilon} \rangle = \mathbf{E}$, and thus $\Delta \bar{\boldsymbol{\sigma}} = \mathbf{0}$ and $\Delta \bar{\boldsymbol{\varepsilon}} = \mathbf{0}$. This is in agreement with the theoretical prediction, cf. Sect. 5.3.1. Finally, we note that the stress and strain fluctuations decay exponentially with the distance from the surface.

Let us now consider the average elastic strain energy density $\bar{w} = \langle w \rangle$, where $w = \frac{1}{2} \boldsymbol{\varepsilon} \cdot \mathbf{L} \boldsymbol{\varepsilon} = \frac{1}{2} \boldsymbol{\sigma} \cdot \boldsymbol{\varepsilon}$. Combining equations (5.60) and (5.61), the average elastic strain energy is obtained in the following form, cf. (5.33),

$$2\bar{w} = 2\langle w \rangle = \langle \boldsymbol{\sigma} \cdot \boldsymbol{\varepsilon} \rangle = \boldsymbol{\Sigma} \cdot \mathbf{E} + \langle \tilde{\boldsymbol{\sigma}} \cdot \tilde{\boldsymbol{\varepsilon}} \rangle \tag{5.62}$$

where the additional energy $\bar{w}^*$ associated with the inhomogeneity of the surface traction is

$$2\bar{w}^* = \langle \tilde{\boldsymbol{\sigma}} \cdot \tilde{\boldsymbol{\varepsilon}} \rangle = \frac{(p^*)^2}{E}(1+\nu)[1 - 2\nu + 2(\alpha y_3)^2]\, e^{-2\alpha y_3}. \tag{5.63}$$

As both the stresses and the strains decay exponentially, the additional energy $\bar{w}^*$ decays even faster – note the terms $e^{-\alpha y_3}$ and $e^{-2\alpha y_3}$ in equations (5.60)–(5.61) and (5.63), respectively. The distribution of the normalized additional energy $\bar{w}^* E/p^{*2}$ for different values of the Poisson's ratio is shown in Fig. 5.3.

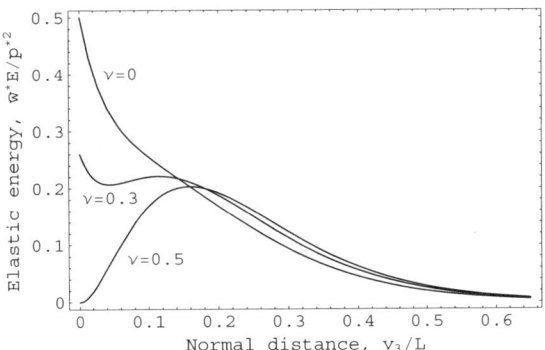

Fig. 5.3. Additional elastic strain energy $\bar{w}^* E/p^{*2}$ in the elastic boundary layer as a function of the distance from the surface y_3/L.

In the elastic case, the solution of the microscopic problem is formed by the superposition of the macroscopic state and the microscopic fluctuations which are independent of the macroscopic state. This, however, is not the case when elasto-plastic deformations are allowed within the boundary layer. In order to illustrate that, let us consider the equivalent Huber-von Mises stress, $\sigma_{\mathrm{eq}} = (\frac{3}{2} \mathbf{s} \cdot \mathbf{s})^{1/2}$, in the elastic boundary layer under consideration. The distribution of the equivalent stress σ_{eq} is shown in Fig. 5.4 for different values of the macroscopic in-plane stress Σ_{11}. In this example, the macroscopic state is specified by $\Sigma_{33} = -\bar{p} = -p^*$ and $\Sigma_{13} = \bar{q} = p^*/2$, where p^* is the fluctuation amplitude, cf. (5.59). It is seen in Fig. 5.4 that the equivalent stress attains its maximum at some distance from the surface, so that plastic yielding would initiate below the surface. This is a well-known result in contact mechanics, cf. Johnson [56]. Secondly, as the in-plane stress Σ_{11} increases, the equivalent stress σ_{eq} increases in the whole domain and, importantly, the position of the maximum changes. Thus, elasto-plastic deformations within boundary layers are expected to be significantly affected by macroscopic stresses. This is discussed more in the next chapter, where contact boundary layers in elastic-plastic solids are analyzed using the finite element method.

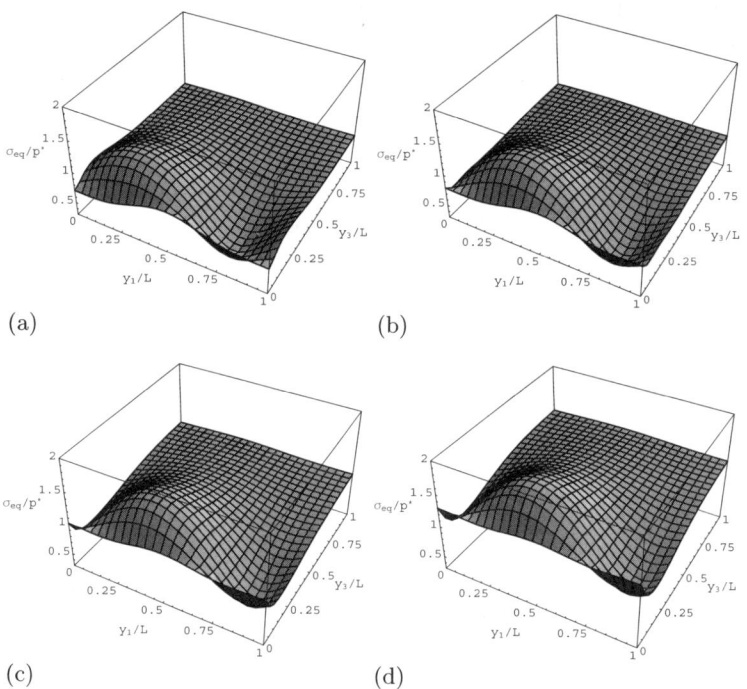

Fig. 5.4. Distribution of the equivalent stress σ_{eq}/p^* in the elastic boundary layer: (a) $\Sigma_{11} = -p^*$, (b) $\Sigma_{11} = -p^*/2$, (c) $\Sigma_{11} = 0$, (d) $\Sigma_{11} = p^*/2$.

5.5 Conclusions

A micromechanical framework has been developed for the analysis of deformation inhomogeneities within the boundary layers induced by micro-inhomogeneous boundary conditions. An averaging operation has been introduced which averages the inhomogeneities along the surface, but preserves the dependence of the respective average quantities on the distance from the surface. The averaging operation decomposes thus an arbitrary field within the boundary layer into its average and fluctuation, the former depends only on the coordinate normal to the surface.

Several properties of such averages have been derived, the most important are recalled below. First of all, the average strain $\bar{\varepsilon}(y_3)$ and the average stress $\bar{\sigma}(y_3)$ satisfy the compatibility conditions analogous to those holding locally at a bonded interface, cf. Sect. 2.4. The corresponding changes with respect to the macroscopic quantities, $\Delta\bar{\varepsilon}$ and $\Delta\bar{\sigma}$, are thus orthogonal. It has also been shown that the average work is not equal to the work of average stress and strain. These properties hold for any constitutive relations describing the material behaviour.

Further, the interfacial relationships have been derived for the case of homogeneous elasticity with, possibly inhomogeneous, plastic strain (eigen-strain). These relationships resemble those holding at bonded interfaces, cf. Sect. 2.5. Finally, it has been shown that an elasto-plastic (or rigid-plastic) boundary layer is macroscopically weakened due to inhomogeneity of deformation.

6

Finite Element Analysis of Contact Boundary Layers

Abstract: This chapter is devoted to the analysis of the boundary layers induced by contact of rough bodies. The implementation issues related to finite element modelling of contact boundary layers are first discussed. Next, two representative asperity interaction problems in elastic-plastic solids are studied. Attention is paid to the interaction of the inhomogeneous boundary layer fields with the macroscopic deformation and to the related effects on the macroscopic contact response.

6.1 Introduction

In this chapter, the general framework developed in Chaps. 4 and 5 is applied to study representative asperity interaction problems in elasto-plastic solids. As in many other application areas, numerical methods, such as the finite element method, are necessary to obtain solutions for practical problems of contact boundary layers. Selected implementation aspects are thus discussed below, that are related to the application of the finite element method to the boundary value problems of contact boundary layers.

Two numerical examples are also provided. Firstly, asperity ploughing is modelled by considering the sliding contact of an initially smooth elasto-plastic body with a rough and rigid obstacle. Next, the contact response of a rough surface compressed by a flat and rigid counter-surface is studied. In the latter case, the analysis is carried out for a real three-dimensional topography of a sand-blasted surface. In both numerical examples, special attention is paid to the effects of the macroscopic in-plane strain on the macroscopic contact response. As discussed in Sect. 3.3, such effects are essential in metal forming processes in which the macroscopic deformations are plastic. The analysis of this chapter is focused on asperity interaction in the elasto-plastic regime. The related effects seem not to have been discussed in the literature yet.

Development of a computational model of a unit cell of the boundary layer involves specification of the material behaviour within the layer and of the boundary conditions for the unit cell, including contact interactions on the surface. Application of the finite deformation framework instead of the infinitesimal strain framework which was used in Chaps. 4 and 5 is thus straightforward: appropriate material model must be incorporated and the boundary conditions remain essentially unchanged with $\mathbf{H}$ in (5.10) being now the macroscopic deformation gradient, see also (6.7) and (6.8). The finite deformation framework, which is more relevant in practical applications, is thus used in the numerical examples presented in this chapter.

The finite element computations reported in this chapter have been performed within the *Computational Templates* environment, and the symbolic code generation system *AceGen* has been used to generate the necessary finite element codes, cf. Korelc [66, 67, 68].

6.2 Remarks on Finite Element Implementation

6.2.1 Scaling

In Chaps. 4 and 5, the equations of the microscopic problem, governing the deformation in the boundary layer, are formulated in the scaled spatial variable $\mathbf{y} = \mathbf{x}/\varepsilon$, where $\varepsilon \ll 1$ is a small parameter expressing the ratio of characteristic dimensions at the micro- and macro-scale. The displacement correction $\mathbf{w}(\mathbf{y})$, being the basic unknown in the microscopic problem, is also a scaled physical displacement, with the same scaling factor ε. At the same time, the microscopic strains and stresses involved in the description are the physical quantities.

The asymptotic analysis is a convenient and rigorous technique to derive the boundary layer equations, and scaling is an important element of this technique. However, in practical applications, it is more convenient to carry out the finite element analysis, or another method of solution of the microscopic problem, using the non-scaled (macroscopic) spatial variable $\mathbf{x} = \varepsilon \mathbf{y}$ and the physical displacement $\varepsilon \mathbf{w}(\mathbf{x}/\varepsilon)$ as the basic unknown. The corresponding governing equations can be obtained by simply putting $\varepsilon = 1$. As ε does not appear in the governing equations, see Sects. 5.2 and 5.3.1, the equations do not change, while the spatial variable $\mathbf{y}$ and the displacement correction $\mathbf{w}(\mathbf{y})$ become physical quantities. This convention is used throughout the present chapter.

6.2.2 Truncated Strip V_H

The spatial domain of the boundary layer problem is not bounded, since the strip V, which constitutes the unit cell, is half-infinite, i.e. $y_3 \in (0, \infty)$, cf. Sect. 5.1. This difficulty can be overcome by considering a *truncated* strip V_H

with $y_3 \in (0, H)$, where H is chosen sufficiently large for the results not to be affected by the truncation (with a desired accuracy). For elliptic problems, for which the Saint-Venant's principle holds, this presents no difficulty and H can be chosen by numerical experiments.

Consider, for example, the problem analyzed in the previous section, i.e. an elastic half-space subjected to sinusoidal traction. The effect of the height H of the truncated strip V_H on the maximum vertical displacement u_{max} is illustrated in Fig. 6.1 for two densities of a regular finite element mesh ($N_{el} = 20$ or $N_{el} = 40$ elements along the boundary). The maximum displacement is normalized by the theoretical value $u_{th} = (1 - \nu^2)p^*l/(\pi E)$, cf. (5.58). It follows from the analytical solution for a half-space that the stresses and strains decay very quickly (exponentially) with the distance from the surface, cf. Sect. 5.4. Accordingly, in the present case, the maximum displacement u_{max} increases by less than 0.1 per cent as the strip height is increased from $H/l = 1$ to $H/l = 2$, where l is the period of sinusoidal traction. In the elastic case, the height of the truncated strip $H = l$ should thus be sufficient for most purposes.

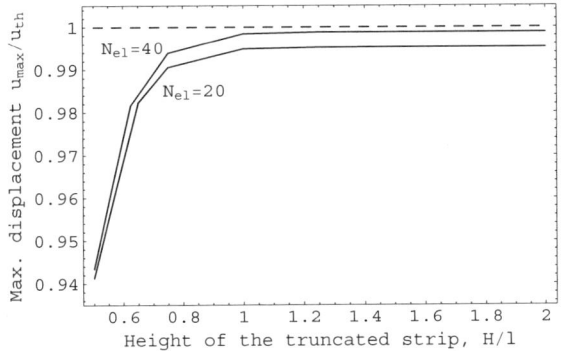

Fig. 6.1. Elastic half-space subjected to sinusoidal traction: effect of the truncation height H on the maximum displacement u_{max}.

The notion of kinematic admissibility must be modified if the strip V is replaced by the truncated strip V_H. One possibility is to require the admissible correction $\mathbf{w}(\mathbf{y})$ to vanish at $y_3 = H$,

$$\mathbf{w}(\mathbf{y}) = \mathbf{0} \quad \text{for} \quad y_3 = H. \tag{6.1}$$

This condition replaces the condition of vanishing gradient of $\mathbf{w}(\mathbf{y})$ far from the boundary, cf. (5.13). The choice of the truncation height H determines the accuracy with which the latter condition is approximated.

Once condition (6.1) is adopted, the solution of the boundary layer problem is unique, while, in the case of the original boundary layer problem with inhomogeneous surface traction, the solution is defined up to an additive vector, cf. Chap. 4. On the other hand, as discussed below, adopting condition (6.1) may require special treatment of the microscopic contact conditions.

6.2.3 Choice of the Basic Unknown

The boundary layer displacement correction $\mathbf{w}(\mathbf{y})$ is the basic unknown of the microscopic problem, cf. Chaps. 4 and 5. However, the displacement correction $\mathbf{w}(\mathbf{y})$ and the total displacement $\mathbf{u}(\mathbf{y})$ are directly related by (5.10). Thus, in practice, either of the two can be adopted as a basic unknown, as long as the constitutive relations and boundary conditions are consistently formulated.

Assume first that the displacement correction $\mathbf{w}(\mathbf{y})$ is chosen as the basic unknown in the finite element formulation. The weak form of the equilibrium equation is obtained by multiplying the local equilibrium equation (5.21) by a kinematically admissible test function $\delta\mathbf{w}(\mathbf{y})$, by integrating over the truncated strip V_H, and by applying the divergence theorem. This standard procedure yields

$$\int_{V_H} \boldsymbol{\sigma} \cdot \mathbf{e}(\delta\mathbf{w}) \, \mathrm{d}V = \int_{\partial V_H} (\boldsymbol{\sigma}\mathbf{n}') \cdot \delta\mathbf{w} \, \mathrm{d}S. \tag{6.2}$$

Here, the stress is related to the unknown displacement correction $\mathbf{w}$ by the constitutive relation, $\boldsymbol{\sigma} = \boldsymbol{\sigma}(\mathbf{E} + \mathbf{e}(\mathbf{w}))$. Since, by assumption, $\mathbf{w}$ is a kinematically admissible correction, it is V-periodic. This implies that the stress $\boldsymbol{\sigma}$ is also V-periodic, in agreement with the requirement of static admissibility of $\boldsymbol{\sigma}$.

Consider the integral on the right-hand side of the weak form (6.2). In view of V-periodicity of $\delta\mathbf{w}$ and anti-periodicity of $\boldsymbol{\sigma}\mathbf{n}'$, the contributions from the lateral faces of V_H cancel each other. Also, the integral over the boundary $y_3 = H$ vanishes since $\delta\mathbf{w} = \mathbf{0}$ at $y_3 = H$, cf. the boundary condition (6.1). As a result, the *principle of virtual work* takes the following form,

$$\int_{V_H} \boldsymbol{\sigma}(\mathbf{E} + \mathbf{e}(\mathbf{w})) \cdot \mathbf{e}(\delta\mathbf{w}) \, \mathrm{d}V = \int_{S_r} \mathbf{t} \cdot \delta\mathbf{w} \, \mathrm{d}S, \tag{6.3}$$

where $\mathbf{t}$ is the traction on surface S_r, resulting either from the traction boundary condition (then $\mathbf{t} = \mathbf{t}^b$) or from the contact interaction. In the case of the displacement boundary condition prescribed on S_r, the integral on the right-hand side of (6.3) vanishes since then $\delta\mathbf{w} = \mathbf{0}$ on S_r.

In the weak form (6.3), both the unknown displacement correction $\mathbf{w}$ and the test function $\delta\mathbf{w}$ are kinematically admissible, which implies their V-periodicity. This condition can be released by introducing the periodicity condition into the weak form using the Lagrange multiplier technique. This leads to the following weak form

$$\int_{V_H} \boldsymbol{\sigma}(\mathbf{E} + \mathbf{e}(\mathbf{w})) \cdot \mathbf{e}(\delta\mathbf{w}) \, \mathrm{d}V = \int_{S_r} \mathbf{t} \cdot \delta\mathbf{w} \, \mathrm{d}S$$
$$+ \int_{\partial V_H^+} \boldsymbol{\lambda} \cdot (\delta\mathbf{w}^+ - \delta\mathbf{w}^-) \, \mathrm{d}S, \tag{6.4}$$

$$\int_{\partial V_H^+} \delta\boldsymbol{\lambda} \cdot (\mathbf{w}^+ - \mathbf{w}^-) \, dS = 0, \tag{6.5}$$

in which $\mathbf{w}$ and $\delta\mathbf{w}$ are not required to be V-periodic. Here, $\mathbf{w}^\pm = \mathbf{w}(\mathbf{y}^\pm)$, $\mathbf{y}^\pm \in \partial V_H^\pm$, and $\boldsymbol{\lambda}$ is a field of Lagrange multipliers defined on ∂V_H^+.

Finite element equations can now be derived from the weak form (6.4)–(6.5) by introducing finite element interpolations of the displacement correction $\mathbf{w}$ and Lagrange multiplier field $\boldsymbol{\lambda}$. This procedure is standard and is omitted here, e.g. Zienkiewicz and Taylor [169]. The only non-standard requirement is that the constitutive relation $\boldsymbol{\sigma} = \boldsymbol{\sigma}(\mathbf{E} + \mathbf{e}(\mathbf{w}))$ involves the macroscopic strain $\mathbf{E}$, as a given data, in addition to the strain derived from the displacement correction $\mathbf{w}$. Thus dedicated finite elements must be used to implement such constitutive relation.

This can be avoided by choosing, as a basic unknown, the total displacement $\mathbf{u}$, which is related to $\mathbf{w}$ by (5.10). Accordingly, the boundary condition (6.1) and V-periodicity of $\mathbf{w}$ must be expressed in terms of $\mathbf{u}$. Noting that $\delta\mathbf{w} = \delta\mathbf{u}$, the weak form (6.4)–(6.5) can be rewritten to yield

$$\int_{V_H} \boldsymbol{\sigma}(\mathbf{e}(\mathbf{u})) \cdot \mathbf{e}(\delta\mathbf{u}) \, dV = \int_{S_r} \mathbf{t} \cdot \delta\mathbf{u} \, dS$$
$$+ \int_{\partial V_H^+} \boldsymbol{\lambda} \cdot (\delta\mathbf{u}^+ - \delta\mathbf{u}^-) \, dS, \tag{6.6}$$

$$\int_{\partial V_H^+} \delta\boldsymbol{\lambda} \cdot [\mathbf{u}^+ - \mathbf{u}^- - \mathbf{H}(\mathbf{y}^+ - \mathbf{y}^-)] \, dS = 0, \tag{6.7}$$

where the boundary condition (6.1) takes now the form

$$\mathbf{u} - \mathbf{u}_0 - \mathbf{H}\mathbf{y} = 0, \quad \delta\mathbf{u} = 0 \quad \text{for } y_3 = H. \tag{6.8}$$

We note that the usual constitutive relation $\boldsymbol{\sigma} = \boldsymbol{\sigma}(\mathbf{e}(\mathbf{u}))$ appears in (6.6), instead of $\boldsymbol{\sigma} = \boldsymbol{\sigma}(\mathbf{E}+\mathbf{e}(\mathbf{w}))$ in (6.4). Accordingly, standard solid elements can be used at the cost of the boundary condition (6.8) and the periodicity condition in (6.7) being somewhat more complicated compared to their counterparts expressed in terms of $\mathbf{w}$.

6.2.4 Prescribed Macroscopic Contact Traction

Consider now the boundary layer in the case of contact of a rough body with a rigid and smooth obstacle. The discussion below is restricted to the macroscopic contact zone; the macroscopic separation zone does not require separate treatment.

Denote by $\mathbf{t}(\mathbf{y}) = t_N(\mathbf{y})\mathbf{n} + \mathbf{t}_T(\mathbf{y})$ the local contact traction and by $\mathbf{T} = T_N\mathbf{n} + \mathbf{T}_T$ the macroscopic one. As discussed in Sects. 4.4 and 4.5, the macroscopic contact tractions are the averages of the local ones, cf. (4.75),

$$\mathbf{T} = \langle \mathbf{t} \rangle, \qquad T_N = \langle t_N \rangle, \qquad \mathbf{T}_T = \langle \mathbf{t}_T \rangle. \tag{6.9}$$

At the same time, the macroscopic contact traction is related to the macroscopic stress by

$$\mathbf{T} = \mathbf{\Sigma n}, \qquad T_N = \mathbf{n} \cdot \mathbf{\Sigma n}, \qquad \mathbf{T}_T = (\mathbf{I} - \mathbf{n} \otimes \mathbf{n})\mathbf{\Sigma n}. \qquad (6.10)$$

Since the macroscopic stress $\mathbf{\Sigma}$ constitutes the input of the microscopic problem, (6.9) and (6.10) impose a constraint on the displacement correction $\mathbf{w}$, which otherwise is only defined up to an additive constant. In a general case, this constraint may be inconsistent with the boundary condition (6.1) imposed on the displacement correction in the truncated strip. Possible treatments of this inconsistency are discussed below.

Let us remind that, far from the boundary, the stress $\boldsymbol{\sigma}$ in the boundary layer tends to the macroscopic stress $\mathbf{\Sigma}$. Thus, instead of prescribing the displacement at $y_3 = H$, as in (6.1), a traction boundary condition can be imposed at $y_3 = H$, namely

$$\boldsymbol{\sigma}\mathbf{n} = \mathbf{T} \quad \text{for} \ \ y_3 = H. \qquad (6.11)$$

In fact, this boundary condition is only applicable in the macroscopic sticking zone, where both the normal displacement and the tangential velocity are constrained by the frictional contact conditions, cf. Sect. 4.5.5. On the contrary, in the macroscopic sliding zone, only the normal displacement is constrained, cf. Sect. 4.5.4, thus a mixed boundary condition must be applied, namely

$$\mathbf{n} \cdot \boldsymbol{\sigma}\mathbf{n} = T_N, \quad (\mathbf{I} - \mathbf{n} \otimes \mathbf{n})\mathbf{w} = \mathbf{0} \quad \text{for} \ \ y_3 = H. \qquad (6.12)$$

Boundary conditions (6.11) and (6.12), sketched in Fig. 6.2(a,b), allow using the macroscopic contact traction as a control parameter of the microscopic problem, according to the coupled micro-macro solution scheme outlined in Chap. 4.

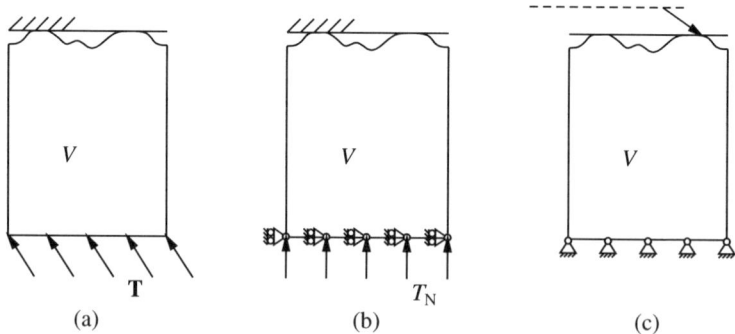

(a)   (b)   (c)

Fig. 6.2. Boundary conditions imposed on the truncated strip: (a) macroscopic sticking zone, cf. (6.11); (b) macroscopic sliding zone, cf. (6.12); (c) alternative control scheme.

However, an alternative approach is also possible, in which the displacement at $y_3 = H$ is prescribed according to (6.1) or (6.8) and the *position* of the rigid obstacle is adopted as a control parameter, cf. Fig. 6.2(c). Note that, so far, it has been assumed that the position of the obstacle is fixed. Thus, for a prescribed displacement of the rigid obstacle, the macroscopic contact traction is obtained as one of the results of the analysis. This scheme cannot be directly applied for the coupled micro-macro problem, however, it allows analysis of a contact boundary layer without referring to any specific macroscopic problem.

6.3 Elasto-Plastic Ploughing

6.3.1 Problem Description

Ploughing is known to be the second, after adhesion, main mechanism of friction, e.g. Bowden and Tabor [16]. Several models of ploughing friction can be found in the literature. Slip-line field theory solutions of plastic flow imposed by a sliding wedge-shaped asperity can be found for example in Challen and Oxley [21], Suh and Sin [144] and Petryk [101]. On the other hand, simple estimates can be obtained by assuming that the average local contact pressure at the asperity contact is equal to the indentation hardness, while the friction stress is related to the average asperity slope angle θ, cf. Fig. 6.3(a). Assuming frictionless conditions at microscopic contacts, i.e. neglecting the adhesive friction component, the macroscopic ploughing friction coefficient is then found from a simple formula

$$\bar{\mu}_{\mathrm{pl}} = \tan\theta. \tag{6.13}$$

Models of this type, applied for different shapes of hard asperities (e.g. wedges, cones, spheres), can be found, for instance, in Rabinowicz [109], Suh and Sin [144], and Komvopoulos et al. [65]. If the asperity slope is not constant, as, for example, in the case of spherical or conical asperities, an average asperity slope can be used in formula (6.13), cf. Fig. 6.3(a).

The models discussed above refer to rigid-plastic materials, the elastic strains are thus neglected. If the elastic strains are significant, i.e. if the so-called rheological factor (see Johnson [55], Bucaille et al. [18]) is small, then

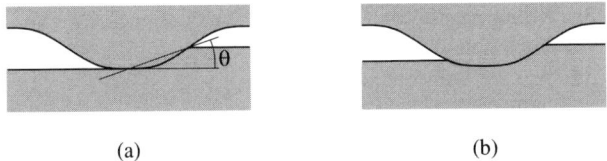

(a)   (b)

Fig. 6.3. Asperity ploughing: (a) average asperity slope in the rigid-plastic case, (b) elastic recovery in the elastic-plastic case.

the ploughing friction coefficient is reduced due to the elastic recovery, as schematically illustrated in Fig. 6.3(b).

The ploughing component of friction is directly related to plastic deformations in the subsurface layer which are induced in a softer body by the asperities of the harder body, or by hard abrasive particles. It can be expected that the interaction of the macroscopic stresses and strains with the localized deformation at the asperity scale may result in the dependence of the macroscopic contact properties (represented, for example, by the macroscopic friction coefficient) on the macroscopic elastic strain. Such effects are observed, when the material deforms plastically at the macro-scale, cf. Sect. 3.3.

The aim of the present numerical example is thus to study the interaction of macroscopic and microscopic deformation fields accompanying elastoplastic ploughing. The related issues have not been addressed in the literature yet. Secondly, the present example illustrates the micromechanical framework developed in Chap. 5. As, in the present chapter, the framework is extended to the case of rough contact interactions, several aspects concerning its application for contact boundary layers are discussed in detail.

Consider thus a periodic array of rigid sine-shaped asperities which plough through an elasto-plastic half-space, cf. Fig. 6.4(a). Plane-strain conditions are assumed which correspond to the case of long asperities aligned perpendicularly to the sliding direction. Clearly, the periodic layout of rigid asperities implies periodicity of the solution in the boundary layer. The corresponding unit cell used in the computations, i.e. the truncated strip V_H, is indicated in Fig. 6.4(a). The body is macroscopically elastic, however, plastic deformations may be induced at the micro-scale, i.e. in the boundary layer, as a result of localized asperity interaction. Accordingly, the elasto-plastic material model with linear hardening is adopted. The geometrical and material parameters used in the simulations are provided in Table 6.1.

The following loading history is assumed. First, the macroscopic in-plane tensile or compressive strain $E_{11} = E_{11}^0$ is applied. At this stage the surfaces are separated, thus the macroscopic normal contact traction is equal to zero,

Table 6.1. Asperity ploughing: material and geometrical parameters.

Parameter	Symbol	Value
Young's modulus	E	71 GPa
Poisson's ratio	ν	0.33
yield stress	σ_y	150 MPa
linear hardening modulus	K	50 MPa
microscopic friction coeff.	μ	0, 0.1
height of the unit cell	H/l	2
asperity height	h/l	0.04
macroscopic strain	E_{11}^0	−0.0015, 0, 0.0015
approach	δ/h	0.025, 0.05, 0.075, 0.1, 0.125, 0.15

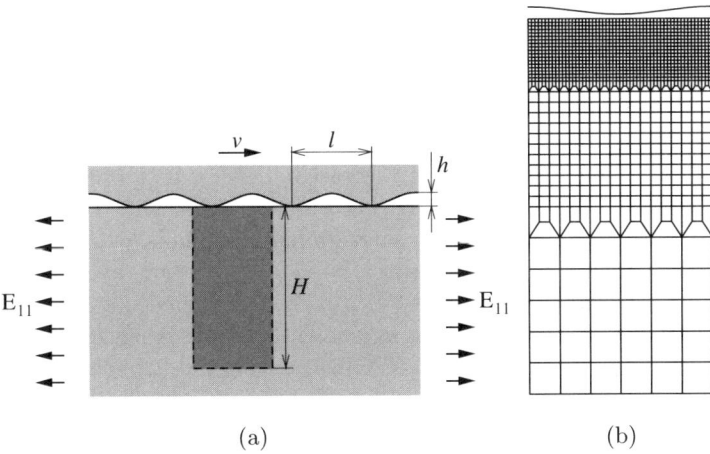

Fig. 6.4. Asperity ploughing: (a) geometry; (b) finite element mesh.

$T_N = \Sigma_{33} = 0$. Secondly, the surfaces are brought into contact, and a relative sliding is imposed, so that the rigid asperities move relative to the surface (to the right in Fig. 6.4). At the same time the nominal separation of the surfaces is decreased from $h/2$ to $h/2 - \delta$, where δ is the prescribed relative approach, i.e. the nominal penetration of asperity summits. The prescribed approach δ is attained after sliding distance of one asperity wavelength l. Subsequently, the sliding distance of $2l$ is imposed at the constant approach δ. Finally, the surfaces are separated, and the macroscopic in-plane strain is released to $E_{11} = 0$. Below, the results of computations are provided for three values of the macroscopic strain E_{11} and for six values of the approach δ/h, cf. Table 6.1.

Two cases are considered regarding the microscopic contact conditions: frictionless contact and Coulomb friction with a local friction coefficient $\mu = 0.1$. The impenetrability and friction conditions are enforced nodally using the augmented Lagrangian technique, cf. Pietrzak and Curnier [106]. The exact contact kinematics is adopted based on the closest point projection of the nodes onto the rigid surface, see, for instance, Pietrzak and Curnier [106]. Accordingly, the exact finite strain kinematics is used to describe the elasto-plastic deformations within the boundary layer. The finite strain framework is preferable also in view of significant configuration changes induced by repeated asperity ploughing.

The finite-strain multiplicative J_2 elasto-plastic model is adopted as a constitutive model of the solid, cf. Simo and Hughes [127]. The incremental formulation assumes hyperelastic isotropic response relative to the local intermediate configuration with the inverse plastic right Cauchy-Green tensor as a state variable. Details concerning the finite element implementation can be found in Stupkiewicz et al. [133], in the Appendix.

On the other hand, only the small-strain formulation is consistent with the micromechanical framework developed in Chap. 5. The computations have thus been repeated adopting the small-strain formulation, and the results have not been found to differ significantly. Unless stated otherwise, the results presented below correspond to the finite-strain case.

The finite element mesh of the truncated strip V_H is shown in Fig. 6.4(b). A volumetric-locking-free quadrilateral element employing the volumetric-deviatoric split and Taylor expansion of shape function is used in the computations, cf. Korelc [67]. According to the loading history defined above, the position of the rigid surface is used as a control parameter, and the displacements are prescribed at $y_3 = H$ in agreement with the scheme of Fig. 6.2(c).

The distribution of the plastic multiplier during sliding is shown in Fig. 6.5 at three values of the macroscopic sliding distance, $g_T = l, 2l, 3l$. It is seen that the plastic deformation is localized in the vicinity of the asperity contact, and thus it is inhomogeneous. In fact, it is only because of the localized asperity interaction that plastic deformations appear in the sub-surface layer. Application of an equivalent, in terms of the average value, uniform normal and tangential traction would result in a purely elastic response, as it is the case far from the surface where the inhomogeneities vanish. Accumulation of the permanent shear deformation in the boundary layer, due to repeated ploughing by subsequent asperities, is also clearly seen in Fig. 6.5.

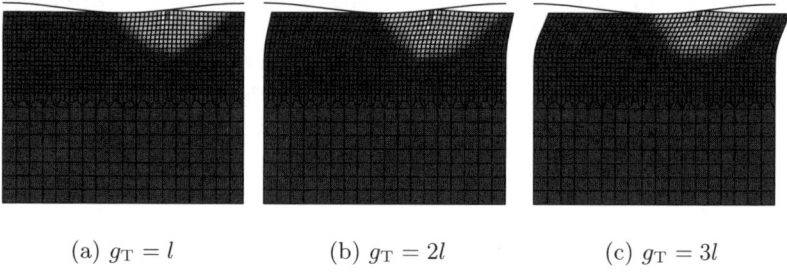

(a) $g_T = l$ (b) $g_T = 2l$ (c) $g_T = 3l$

Fig. 6.5. Asperity ploughing: distribution of plastic multiplier at three values of the macroscopic sliding distance g_T ($\delta/h = 0.15$, $E_{11} = 0$).

6.3.2 Ploughing Friction Coefficient

The macroscopic normal pressure and the macroscopic friction stress have been obtained from the nodal reaction forces at the bottom edge of the unit cell. Due to global equilibrium of the unit cell this is equivalent to averaging the microscopic contact tractions. The macroscopic contact stresses $-T_N$ and T_T, normalized by the yield stress σ_y, are shown in Fig. 6.6 and 6.7 as a function of the sliding distance g_T. Figures 6.6(a) and 6.7(a) present the raw

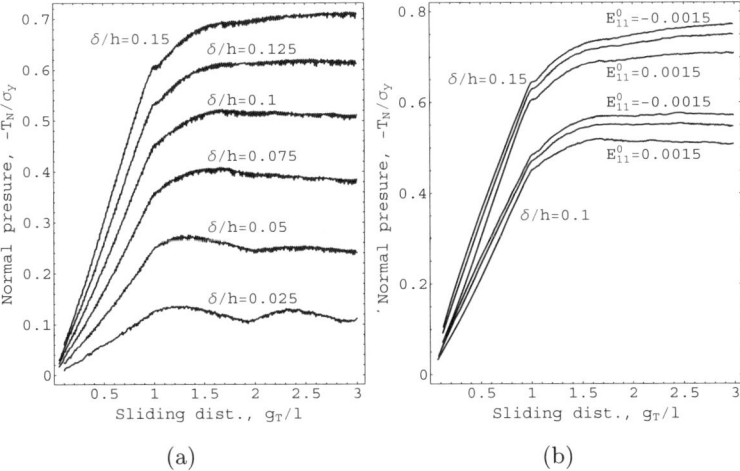

Fig. 6.6. History of the macroscopic contact pressure $-T_N/\sigma_y$: (a) raw data (for $E_{11}^0 = 0.0015$); (b) smoothed data (for $\delta/h = 0.1$ and 0.15).

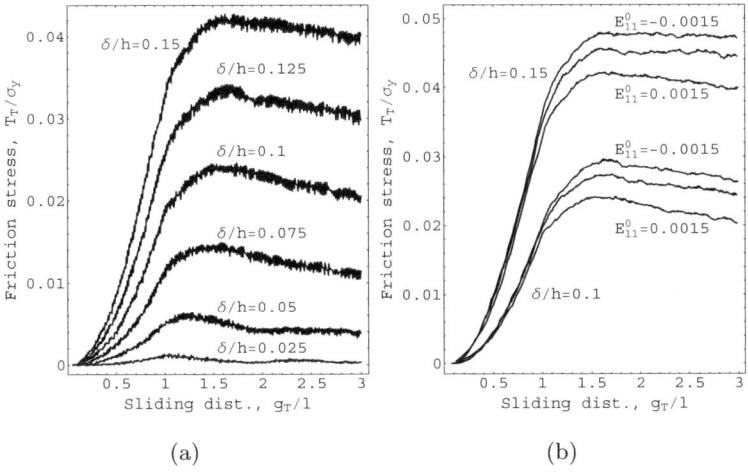

Fig. 6.7. History of the macroscopic friction stress T_T/σ_y: (a) raw data (for $E_{11}^0 = 0.0015$); (b) smoothed data (for $\delta/h = 0.1$ and 0.15).

data obtained for $E_{11}^0 = 0.0015$ (macroscopic in-plane tension) and for $\mu = 0$; the six curves correspond to the six values of the approach δ/h.

Application of boundary conditions of the type shown in Fig. 6.2(c), implies that the approach is controlled rather than the macroscopic normal pressure, the latter resulting from the solution. In particular, the macroscopic normal pressure is not necessarily constant even if the approach is held constant. It is seen in Figures 6.6 and 6.7 that the macroscopic contact tractions

oscillate with respect to the slowly varying average tractions. These oscillations are caused by the finite element discretization and nodal enforcement
of contact conditions: the contact nodes come into contact and loose contact
which results in oscillatory response. These purely numerical oscillations can
be smoothed, for example by local averaging in time. Figures 6.6(b) and 6.7(b)
present the history of smoothed macroscopic contact tractions for two sample
values of the approach δ/h and for three values of the macroscopic strain E_{11}^0.

Figure 6.8 presents the history of the instantaneous macroscopic friction
coefficient $\bar{\mu} = T_T/|T_N|$. The data corresponds to the macroscopic contact
tractions presented in Figures 6.6 and 6.7. It is seen that the macroscopic
friction coefficient, just like the macroscopic contact tractions, is not constant.
In order to estimate the effective ploughing friction coefficient $\bar{\mu}_{pl}$, the time-
averaged coefficient has been determined using the data corresponding to
$2 \leq g_T/l \leq 3$, i.e. by neglecting the running-in period $0 \leq g_T/l < 2$, cf. the
shaded region in Fig. 6.8(a). The ploughing friction coefficient has thus been
determined according to

$$\bar{\mu}_{pl} = \frac{1}{l} \int_{2l}^{3l} \frac{T_T(g_T)}{|T_N(g_T)|} \, \mathrm{d}g_T - \mu. \tag{6.14}$$

In (6.14), the microscopic friction coefficient μ has been subtracted from the
macroscopic one, so that $\bar{\mu}_{pl}$ describes only the ploughing contribution.

The macroscopic ploughing friction coefficient $\bar{\mu}_{pl}$ is shown in Fig. 6.9(a)
as a function of the macroscopic contact pressure $-T_N$. The ploughing friction
coefficient appears to be insensitive to the macroscopic strain E_{11}^0: the three
curves corresponding to three different values of E_{11}^0 practically coincide both

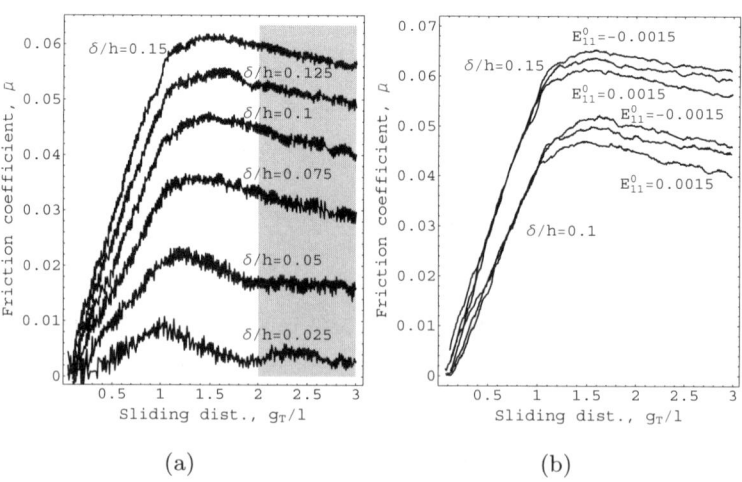

Fig. 6.8. History of the macroscopic friction coefficient $\bar{\mu} = T_T/|T_N|$: (a) raw data
(for $E_{11}^0 = 0.0015$); (b) smoothed data (for $\delta/h = 0.1$ and 0.15).

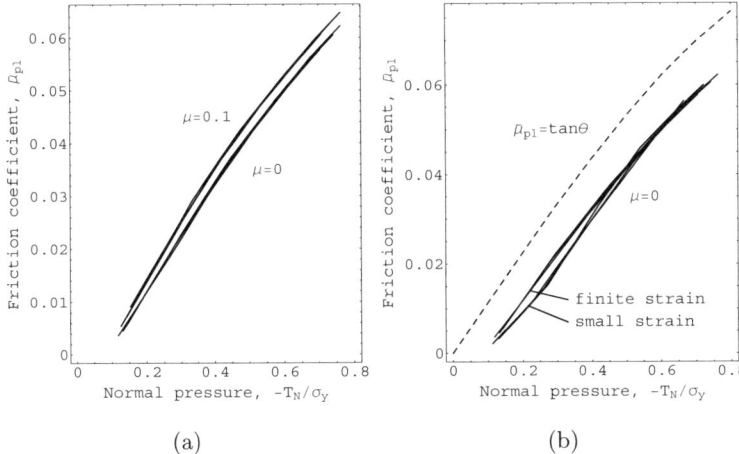

(a)   (b)

Fig. 6.9. Macroscopic ploughing friction coefficient $\bar{\mu}_{\mathrm{pl}}$ as a function of the normalized macroscopic contact pressure $-T_{\mathrm{N}}/\sigma_{\mathrm{y}}$. The curves corresponding to different
values of the macroscopic strain E_{11}^0 practically coincide for fixed microscopic friction coefficient μ, cf. figure (a). Prediction for $\mu = 0$ is compared to the simple model
(6.13) in figure (b).

for $\mu = 0$ and for $\mu = 0.1$. At the same time, the effect of the microscopic
friction coefficient μ on the macroscopic ploughing friction coefficient $\bar{\mu}_{\mathrm{pl}}$ is
clearly visible.

In Fig. 6.9(b), the ploughing friction coefficient $\bar{\mu}_{\mathrm{pl}}$ resulting from the
present analysis is compared to the one predicted by the simple model (6.13).
The average asperity slope angle θ has been determined according to the
scheme of Fig. 6.3(a), while the real contact area has been determined from
the macroscopic contact pressure by assuming indentation hardness $H = 6k$,
where $k = \sigma_{\mathrm{y}}/\sqrt{3}$ is the yield stress in shear. The difference between the
two predictions is approximately constant (i.e. independent of the normal
pressure) and probably results from the elastic recovery, cf. Fig. 6.3(b).

The obtained result that the ploughing friction coefficient $\bar{\mu}_{\mathrm{pl}}$ is not affected by the macroscopic strain E_{11}^0 seems quite surprising in view of the
substantial effect of E_{11}^0 on the relation between the approach δ/h and the
macroscopic normal pressure $-T_{\mathrm{N}}/\sigma_{\mathrm{y}}$. The latter effect is clearly visible in
Fig. 6.6(b) and also in Fig. 6.10: at constant pressure, the approach increases
for macroscopic tension ($\mathrm{E}_{11}^0 > 0$) and decreases for macroscopic compression
($\mathrm{E}_{11}^0 < 0$). However, the response in terms of the macroscopic friction stress is
similar, so that the resulting friction coefficient is not sensitive to macroscopic
straining, at least in the conditions adopted in present computations. Analysis
of the residual stresses in the boundary layer provides a possible explanation
of this behaviour.

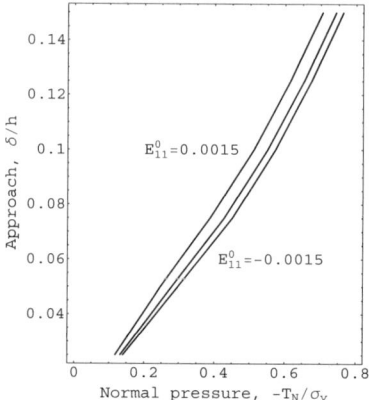

Fig. 6.10. Approach δ/h as a function of the normalized macroscopic contact pressure $-T_N/\sigma_y$ for $\mu = 0$.

6.3.3 Residual Stresses

The distribution of the average in-plane stresses[1] $\bar{\sigma}_{11}$ and $\bar{\sigma}_{22}$ is shown in Fig. 6.11. Here, in order to be consistent with the micromechanical framework of Chap. 5, the small strain formulation has been used. Note that the ploughing friction coefficient is practically not affected by the choice of formulation, cf. Fig. 6.9(b), so that the analysis below should be relevant also for the finite-strain case.

It is seen in Fig. 6.11 that in the vicinity of the surface, say for $y_3/l < 0.2$ in the present case of $\delta/h = 0.1$, the average stresses $\bar{\sigma}_{11}$ and $\bar{\sigma}_{22}$ are not affected by the macroscopic strain E_{11}^0. This suggests that residual stresses develop in the boundary layer, which compensate the macroscopic stresses, so that the stress state in a thin sub-surface layer is practically independent of the macroscopic state far from the surface. Clearly, these residual stresses strongly depend on the macroscopic strain E_{11}^0. This is illustrated in Fig. 6.12 which presents the average stresses $\bar{\sigma}_{11}$ and $\bar{\sigma}_{22}$ after the surfaces are separated and the in-plane strain is released, $E_{11} = 0$, so that the macroscopic stresses are equal to zero.

Let us finally note that the interior part of the average stress $\bar{\sigma}$ (e.g. the $\bar{\sigma}_{11}$ component) deviates from its macroscopic counterpart only in a part of the boundary layer adjacent to the surface, namely in the zone of non-zero plastic deformations. This is in agreement with the interfacial relationships for averages, specified by (5.49).

---

[1] The averages of the other stress components, namely $\bar{\sigma}_{33}$ and $\bar{\sigma}_{13}$, are equal to the macroscopic counterparts in agrement with the compatibility condition $(5.29)_2$.

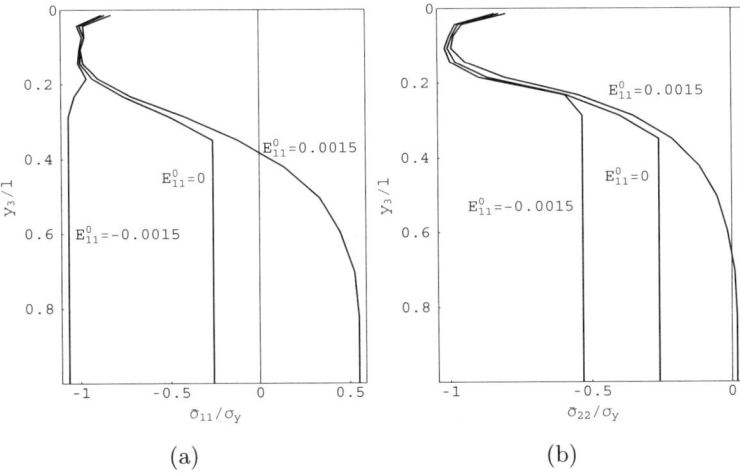

Fig. 6.11. Distribution of average in-plane stresses $\bar{\sigma}_{11}$ (a) and $\bar{\sigma}_{22}$ (b) after sliding distance $g_T = 3l$ (for $\delta/h = 0.1$).

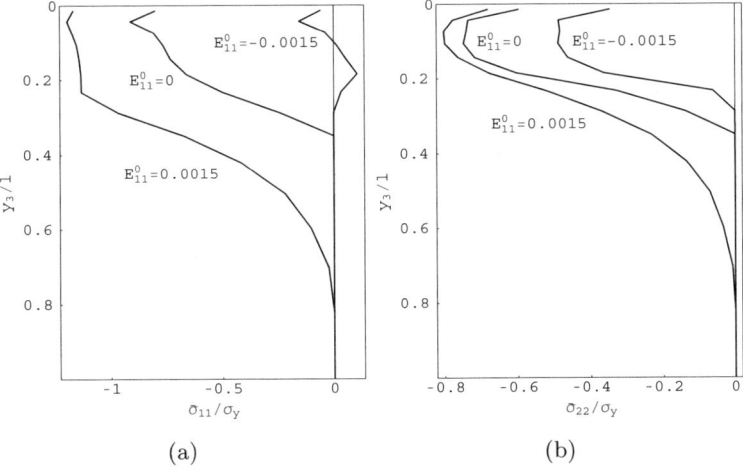

Fig. 6.12. Distribution of average in-plane stresses $\bar{\sigma}_{11}$ (a) and $\bar{\sigma}_{22}$ (b) after sliding distance $g_T = 3l$ and after the in-plane strain is released.

6.4 Normal Contact Compliance of a Sand-Blasted Surface

In this section, the elasto-plastic normal contact compliance of a real three-dimensional rough surface is analyzed using the finite element method. In accord with the main interest of the present chapter, the aim of this example is to study the interaction of the local elasto-plastic deformations, associated with flattening of asperities, with the macroscopic deformation field, and, specifically, to investigate the effect of the macroscopic strain on the normal contact compliance.

The present direct approach to modelling of contact interactions of rough bodies is, in essence, similar to numerous previous studies, e.g. Bandeira et al. [9], Varadi et al. [150]. The original contribution of the present study is that, here, the macroscopic strain and its effect on the contact response are directly accounted for and thoroughly analyzed.

In brief, the present approach involves the following steps. First, a three-dimensional topography of a real surface is measured using the scanning stylus profilometry. A representative part of the surface is next chosen, and a finite element model of the unit cell of the corresponding boundary layer is generated. A boundary value problem is then solved by considering the frictionless contact of the rough surface with a rigid and smooth counter-surface. In addition to all the microscopic quantities, such as local displacements, stresses, etc., the analysis provides the macroscopic contact pressure and the real contact area fraction as a function of the relative approach of the surfaces. Below, several details of the present micromechanical scheme are commented, and the effect of the macroscopic in-plane strain on the macroscopic contact response is studied.

Dedicated measurements of surface roughness and normal contact compliance of a sand-blasted surface, reported in this section, have been performed at the Surface Layer Laboratory of the Institute of Fundamental Technological Problems (IPPT) by Mrs. A. Bartoszewicz, Dr. S. Kucharski, and Dr. G. Starzyński.

6.4.1 Material, Surface Roughness, Finite Element Model

A steel specimen has been sand-blasted in order to create a severely rough surface. Tree-dimensional topography of a $6 \times 6 \, \text{mm}^2$ area has then been measured using a 3D stylus scanning profilometer with the resolution of 2, 20, and 1 micron in the x-, y-, and z-directions, respectively. The following roughness parameters have been determined using this data: the arithmetic mean roughness value $S_a = 6.01 \, \mu\text{m}$, the RMS (root-mean-square) roughness value $S_q = 7.98 \, \mu\text{m}$, and the maximum peak-to-valley height $S_y = 94.0 \, \mu\text{m}$.

The material is a carbon steel C45 (ISO) with the initial yield stress of about 400 MPa and the ultimate tensile strength of about 640 MPa. Although the hardening curve of the bulk material is readily available, strain hardening is neglected in the simulations, and a uniform yield stress $\sigma_y = 400 \, \text{MPa}$ is assumed in the whole volume. This is clearly an approximation since the surface layer is expected to be work-hardened after sand-blasting. However, determination of the actual distribution of the plastic properties and residual stresses within the surface layer after sand-blasting is difficult, and has not been attempted. Thus all the related effects are neglected in the present study.

Considering the typical size and spacing of surface asperities, the whole $6 \times 6\,\text{mm}^2$ area is too large to represent the surface topography with sufficient accuracy after feasible finite element discretization. Clearly, the limitation here is the size of the problem and the computational time needed for the solution of the finite element equations. Thus a $1.08 \times 1.08\,\text{mm}^2$ sub-area has been selected for the subsequent finite element computations. The size of the sample has been chosen arbitrarily, and the problem of optimal choice has not been addressed. However, representativeness of the chosen sample has been verified by comparing the macroscopic contact response of three different samples of the same size. This is discussed later.

Periodicity of roughness layout is an essential assumption of the microme-chanical framework of Chaps. 4 and 5. As the real random surfaces are not periodic, the roughness of the considered sample has been artificially made periodic, so that the micromechanical framework could be consistently applied. A periodic roughness topography has been obtained by modifying the surface in the vicinity of the boundary of the surface sample. The original surface, the modified periodic one, and the difference between the two are shown in Fig. 6.13.

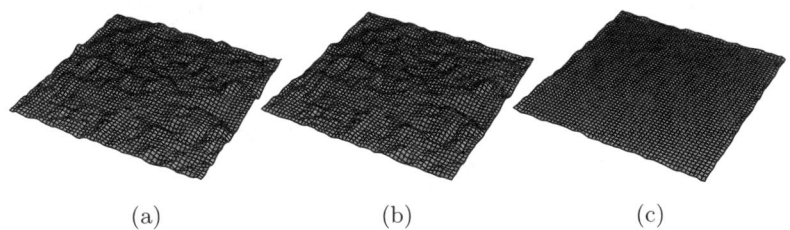

(a)   (b)   (c)

Fig. 6.13. Sand-blasted surface: (a) original $1.08 \times 1.08\,\text{mm}^2$ sample; (b) modified periodic surface; (c) difference between the two.

A structured, three-dimensional finite element mesh has been designed such that the mesh is substantially refined in the vicinity of the surface, see Fig. 6.14. The contact surface is divided into a regular mesh of 54×54 quad-rilateral elements, which gives the size of the sample equal to $1.08 \times 1.08\,\text{mm}^2$ for node spacing of $20\,\mu\text{m}$ (equal to the y-resolution of the scanned profile).

Because the surface is severely rough, flattening of asperities is accompanied by rather large deformations. As the real contact area fraction approaches unity, the equivalent plastic strain reaches 0.9, locally. Accordingly, the finite deformation kinematics is adopted and the finite-strain multiplicative J_2 elasto-plastic model is used, cf. Simo and Hughes [127]. Material and geometrical parameters of the present finite element model are summarized in Table 6.2.

The finite element model shown in Fig. 6.14 consists of 14,888 hexahedral elements. A volumetric-locking-free 8-node element employing the volumetric-

Fig. 6.14. Sand-blasted surface: finite element mesh of the unit cell.

Table 6.2. Sand-blasted surface: material and geometrical parameters.

Parameter	Symbol	Value
Young's modulus	E	210 GPa
Poisson's ratio	ν	0.3
yield stress	σ_y	400 MPa
characteristic length	l	1.08 mm
height of the unit cell	H/l	2
macroscopic strain	E_{11}, E_{22}	−0.001, 0, 0.001

deviatoric split and Taylor expansion of shape function is used, cf. Korelc [67]. The unilateral contact conditions are enforced nodally using the augmented Lagrangian technique, cf. Pietrzak and Curnier [106]. The total number of 52,817 unknowns combined with about 100 time increments per analysis and an average of eight Newton iterations per increment make the present example a moderately large-scale simulation.

6.4.2 Normal Contact Compliance and Representativeness

The unilateral contact condition, specified by the Signorini condition (4.36), applies for ideally smooth surfaces and describes the case of a "rigid" contact interaction: separation occurs if the normal gap is greater than zero, $g_N > 0$; the surfaces are in contact if the gap is equal to zero, $g_N = 0$; and penetration with $g_N < 0$ is ruled out, cf. Fig. 6.15(a).

If, however, the surfaces are rough, then the deformation of the asperities is associated with some equivalent contact compliance. Defining the normal

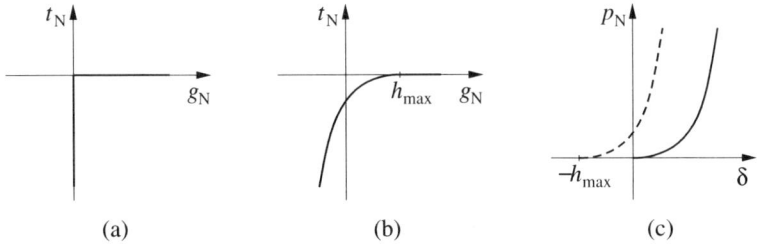

Fig. 6.15. Impenetrability conditions: (a) "rigid" Signorini condition; (b) compliant contact law; (c) compliant contact law expressed in terms of contact pressure $p_N = -T_N$ and approach $\delta = h_{max} - g_N$.

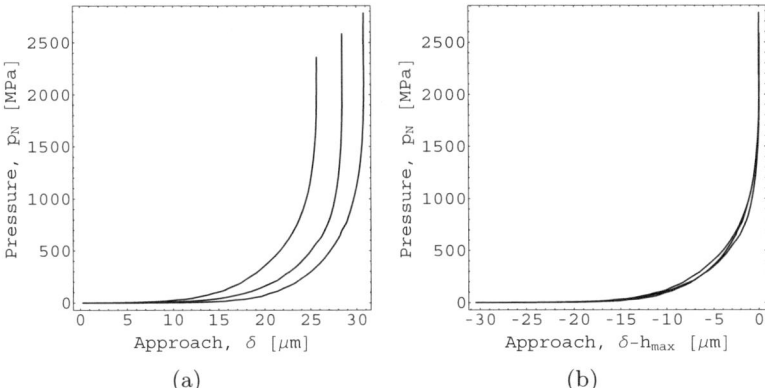

Fig. 6.16. Normal contact compliance predicted for three $1.08 \times 1.08\,\text{mm}^2$ samples: contact pressure p_N as a function of the approach δ (a) and of the corrected approach $\delta - h_{max}$ (b).

gap g_N as the signed distance between the *nominal* surfaces, contact occurs first for $g_N > 0$, i.e. when asperity summits start to interact, and the normal gap decreases as the macroscopic contact pressure $p_N = -T_N$ decreases, cf. Fig. 6.15(b). The related contact compliance is an important factor in many engineering applications.

The compliant contact response can also be expressed in terms of the macroscopic contact pressure $p_N = -T_N$ and approach $\delta = h_{max} - g_N$, cf. the solid line in Fig. 6.15(c). Here, h_{max} is the maximum asperity peak height, measured with respect to the nominal plane, equal to the gap at which contact occurs first.

Three different samples have been simulated in order to check whether the adopted $1.08 \times 1.08\,\text{mm}^2$ sample is representative and sufficiently characterizes the roughness of the analyzed surface. The results are compared in Fig. 6.16. It is seen that the predicted pressure-approach relation, cf. Fig. 6.16(a), is visibly affected by the choice of the sample. This is because the height h_{max} of the highest peak, as a highly random quantity, is different for each sample.

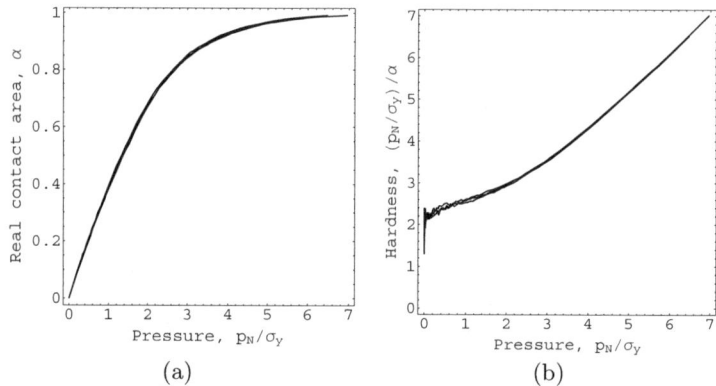

Fig. 6.17. Real contact area fraction (a) and dimensionless hardness (b) predicted for three $1.08 \times 1.08 \, \text{mm}^2$ surface samples.

If, however, h_{max} is subtracted from the approach δ, then the contact response of the three surface elements, expressed in terms of the corrected approach $\delta - h_{\text{max}}$, is practically identical for the three samples, cf. Fig. 6.16(b). Also, the predicted real contact area fraction α and hardness p_N/α, both expressed as a function of the contact pressure p_N normalized by the yield stress σ_y, are almost identical for the three samples, as shown in Fig. 6.17. Concluding, the adopted $1.08 \times 1.08 \, \text{mm}^2$ sample can be considered representative for the whole surface, provided that the approach δ is correctly interpreted.

6.4.3 Experimental Verification of the Finite Element Model

In order to verify the developed finite element model, the normal contact compliance of the sand-blasted surface has been measured experimentally using the technique developed by Handzel-Powierża et al. [38]. The experimental setup is sketched in Fig. 6.18(a). Three hard, smooth, and flat punches of area $8 \times 8 \, \text{mm}^2$ each are pressed into the specimen, and the indentation force as well as the relative approach are measured during loading and unloading. Note that elastic deflections in the experimental setup contribute to the measured approach, as schematically illustrated in Fig. 6.18(b). These deflections are not known.

The following experimental procedure was applied. Three series of complete loading-unloading cycles were performed with the maximum contact pressure p_N^{max} equal to 250, 500, and 750 MPa. In order to increase repeatability of the results, in each case, a prestressing loading-unloading cycle with the maximum pressure of 20 MPa was applied prior to the actual loading cycle.

The recorded pressure-approach diagram corresponding to the maximum pressure of 250 MPa is shown in Fig. 6.19(a) together with the respective diagram predicted by the finite element model. It is evident that the two

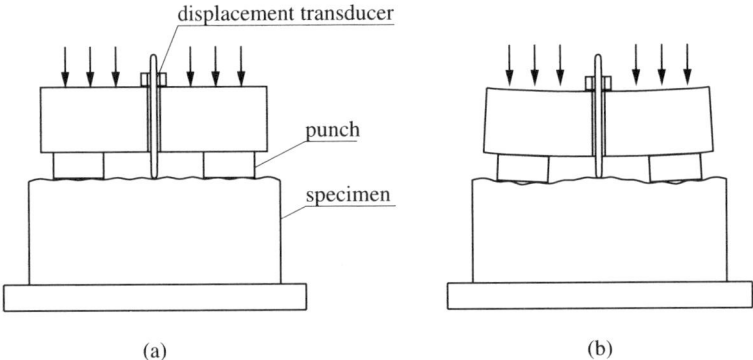

Fig. 6.18. Measurement of the normal contact compliance: (a) scheme of the experimental setup; (b) elastic deflections in the system.

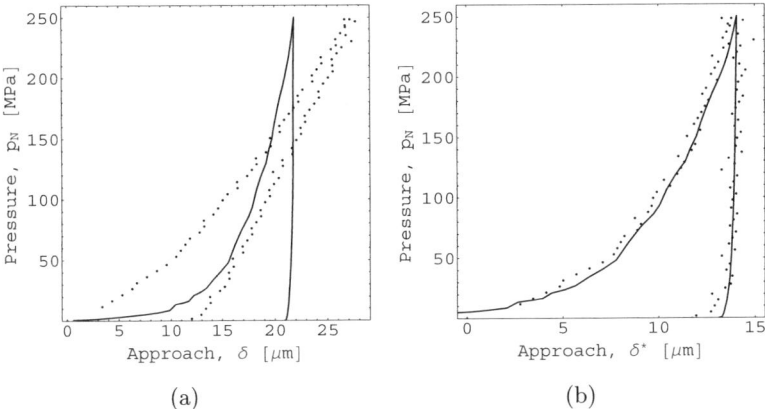

Fig. 6.19. Normal contact compliance of the sand-blasted surface: (a) raw data and (b) data corrected by subtracting the elastic deflections and the initial gap (solid line – finite element prediction, dots – experiment).

diagrams do not match each other. This is, however, not surprising because of the elastic deflections present in the measured approach.

Assuming that the unknown elastic deflections in the experimental setup are proportional to the indentation force, the measured approach δ_{exp} can be expressed as a sum of the actual approach δ^*_{exp}, due to deformation of asperities, and the elastic deflection $k_{\mathrm{exp}}p_{\mathrm{N}}$, proportional to the contact pressure. The approach δ^*_{exp} is thus given by

$$\delta^*_{\mathrm{exp}} = \delta_{\mathrm{exp}} - k_{\mathrm{exp}}p_{\mathrm{N}}, \qquad (6.15)$$

where k_{exp} is an unknown compliance of the experimental setup.

It is also seen in Fig. 6.19(a) that the measured and the predicted residual displacements after complete unloading are different. This is partly related to

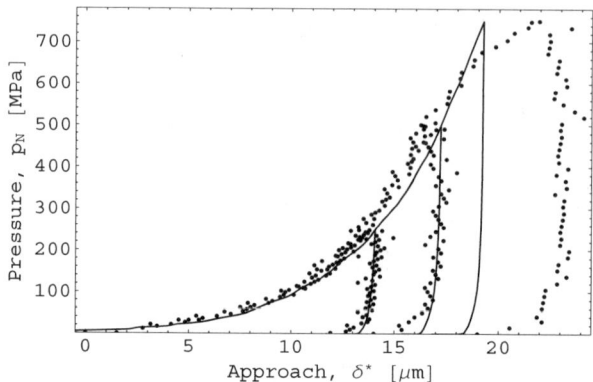

Fig. 6.20. Normal contact response of the sand-blasted surface for three different maximum contact pressures (solid line – finite element prediction, dots – experiment).

a small, but uncontrolled, initial prestressing in the experimental setup. On the other hand, as already discussed, there is quite some ambiguity in determination of the approach predicted by the finite element model, cf. Fig. 6.16. The predicted approach δ_{model} is thus corrected according to

$$\delta^*_{\mathrm{model}} = \delta_{\mathrm{model}} - \delta_0, \qquad (6.16)$$

where δ_0 is an unknown reference approach.

The two unknown parameters in (6.15) and (6.16), k_{exp} and δ_0, have been identified by fitting the *unloading* branch of the pressure-approach curve corresponding to $p_N^{\mathrm{max}} = 250\,\mathrm{MPa}$, and the following values have been found to provide the best fit: $k_{\mathrm{exp}} = 0.0538\,\mu\mathrm{m/MPa}$ and $\delta_0 = 7.79\,\mu\mathrm{m}$. After application of the above correction procedure, a very good agreement of the predicted and the measured contact compliance has been obtained, as shown in Fig. 6.19(b). These parameters have next been used to correct the pressure-approach response corresponding to the two remaining load cases.

The corrected pressure-approach diagrams corresponding to all three loading cases are presented in Fig. 6.20. In the case of $p_N^{\mathrm{max}} = 500\,\mathrm{MPa}$, the obtained agreement is very good for both the loading and the unloading curves. As the case with the highest maximum pressure is concerned, the predicted loading curve follows the experimental one up to the pressure of about 600 MPa. However, with the further increase of the contact pressure, a softening is observed in the experimental response. This is probably due to macroscopic plastic indentation of the punches into the specimen, so that the approach δ^*_{exp} includes displacements due to macroscopic plastic deformations. The experimental unloading curve follows the predicted one, but it is shifted by about $4\,\mu\mathrm{m}$, which is interpreted as the additional displacement due to macroscopic plastic deformations.

Although the predicted and the measured contact response match each other very well, it should be noted that there is still some ambiguity in the finite element model regarding determination of the plastic properties within the surface layer. In fact, the above identification procedure has been repeated assuming the initial yield stress $\sigma_y = 500\,\text{MPa}$ and the linear isotropic hardening modulus $K = 300\,\text{MPa}$, and a very similar matching of the predicted and the measured response has been obtained with parameters $k_{\text{exp}} = 0.0530\,\mu\text{m/MPa}$ and $\delta_0 = 5.89\,\mu\text{m}$ providing the best fit. Note that parameter k_{exp}, i.e. the identified elastic compliance of the experimental setup, appears to be nearly insensitive to the plastic properties of the surface layer. This is because k_{exp} is identified using the response during unloading, which proceeds in an, essentially, elastic regime.

6.4.4 Effect of In-Plane Strain on Contact Response

To conclude this section, we shall study the effect of the macroscopic in-plane strain on the macroscopic contact response. For the study to be more complete, in addition to the real sand-blasted surface studied above, another, relatively smooth surface is also analyzed. In the latter case, an artificial surface topography has been generated by simply scaling the roughness of the sand-blasted surface by the factor of 0.2. Consequently, all the above results regarding the representativeness of the surface sample and the adequacy of the finite element model apply also for the smoother surface. The terms "rough" and "smooth" are used below to denote the two surface topographies.

Similarly to the asperity ploughing example of Sect. 6.3, the contact response is studied here for different macroscopic in-plane strain. In the simulations, the macroscopic strain was enforced before the normal pressure was applied and then it was hold constant during compression. Four cases were considered: (i) $E_{11} = E_{22} = 0$, (ii) $E_{11} = E_{22} = 0.001$, (iii) $E_{11} = E_{22} = -0.001$, (iv) $E_{11} = -E_{22} = 0.001$, where E_{11} and E_{22} denote the in-plane components of the macroscopic strain tensor, and $E_{12} = 0$ was assumed in all cases.

Contact response predicted for cases (i)–(iii) is reported in Figs. 6.21–6.23. Case (iv), i.e. combined tensile-compressive in-plane strain, yielded results which are hardly distinguishable from case (i), and thus the corresponding diagrams are omitted in Figs. 6.21 6.23. This result is in agreement with the model of Giannakopoulos [33], and also with the experimental results of Lee and Kwon [77], who studied the effect of the initial in-plane stresses on the force-depth response of instrumented sharp indentation. They showed that only the average in-plane stress affects the force-depth response whereas the pure shear component, as in the present case (iv), has a negligible effect.

The predicted pressure-approach response is shown in Fig. 6.21. It is seen that application of the macroscopic in-plane tensile strain results, for a fixed contact pressure, in the increase of the approach, and the effect of compressive strain is opposite. Also this result is in agreement with the results of Giannakopoulos [33] and Lee and Kwon [77] concerning the force-depth relationship in indentation.

In Figure 6.22, the pressure-approach diagrams from Fig. 6.21 are compared on a log-log plot. Here, the approach δ is normalized by the arithmetic mean roughness parameter S_a, and the contact pressure p_N is normalized by the yield stress σ_y. Except for very low pressures, the normalized response of the rough surface is very similar to that of the smooth one, the smooth surface being somewhat more compliant. At low contact pressures, the surface with a more rough topography exhibits a significantly stiffer response. Note that the log-log pressure-approach diagrams in Fig. 6.22 are not linear. This indicates that the validity of the popular power law, $p_N = c_N \delta^m$, e.g. Kragelsky et al. [71], Wriggers [165], is rather restricted.

The effect of the macroscopic in-plane strain on the real contact area is illustrated in Fig. 6.23(a). The in-plane tension promotes, and the compression inhibits, asperity flattening, so that, for a fixed contact pressure, the real contact area fraction α is higher in the case of tension, and it is lower in the case of compression. This effect is also clearly visible in Fig. 6.23(b), where the effective hardness, defined as the ratio of the normalized contact pressure p_N/σ_y and the real contact area fraction α, is shown as a function of the normalized contact pressure.

The prediction of the micromechanical model of Wanheim and Bay [10, 156], corresponding to the frictionless case and a small asperity angle, is also included in Fig. 6.23. That prediction is rather close to the results obtained for the rough sand-blasted surface, cf. Fig. 6.23(a), although, for $p_N/\sigma_y < 2$, the hardness predicted by the Wanheim-Bay model is significantly higher than that following from the present modelling, cf. Fig. 6.23(b). This difference can partly be attributed to the elastic deformations which are neglected in the rigid-plastic Wanheim-Bay model. Interestingly, according to the results of

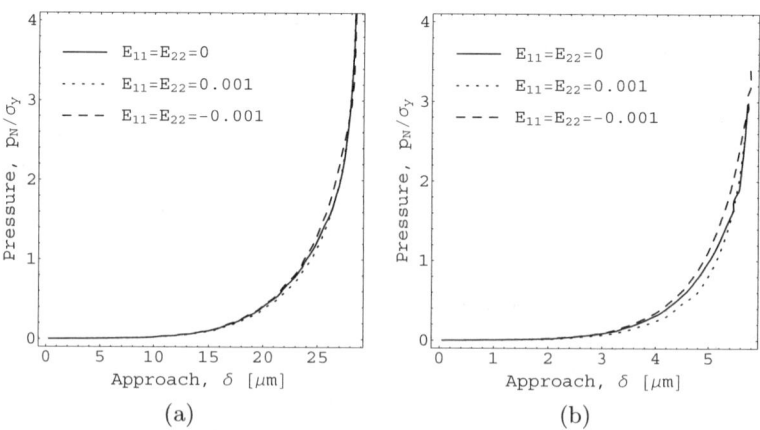

Fig. 6.21. Effect of macroscopic in-plane strain on the normal contact compliance: (a) sand-blasted surface; (b) artificially smoothed surface.

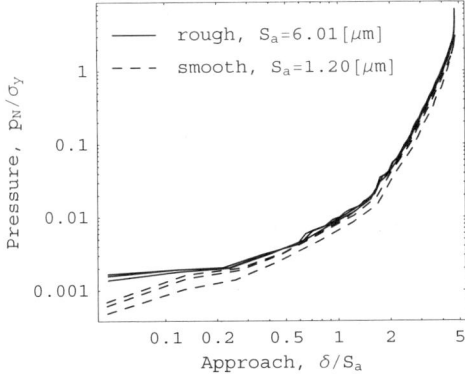

Fig. 6.22. Normalized pressure-approach diagrams on a log-log plot.

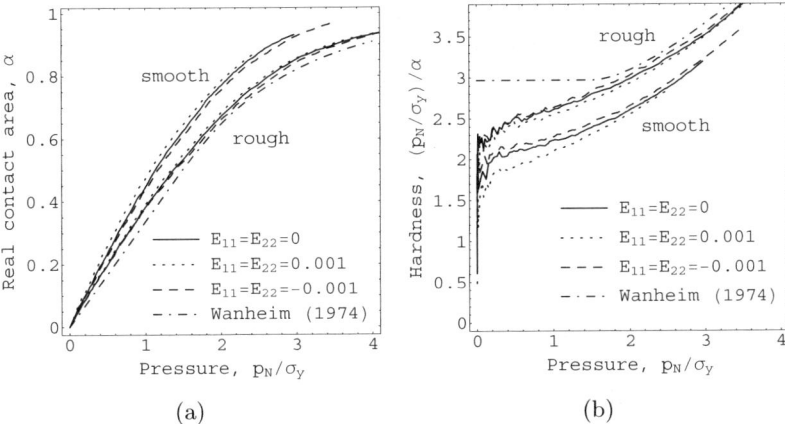

Fig. 6.23. Effect of macroscopic in-plane strain on real contact area (a) and dimensionless hardness (b).

the present elasto-plastic finite-element modelling, the hardness of the rough surface is higher than that of the smooth surface, while an opposite effect is predicted by the slip-line analysis of Bay [10]. It is not clear whether this difference is due to elasticity, due to a completely different surface topology assumed by the two models (note that Bay [10] analyzed flattening of rigid-plastic wedge-like asperities in the plane strain conditions), or due to finite-strain effects and related configurational changes at the asperity scale (which are included in the present modelling).

6.5 Conclusions

Two representative asperity interaction problems of asperity ploughing and flattening in the elasto-plastic regime have been studied in this chapter. The analysis has been focused on the effects of the macroscopic in-plane strains on the contact response. The corresponding predictions are hoped to contribute to a better understanding of the mechanisms of contact interactions of rough bodies. Some remarks concerning the finite element modelling of contact boundary layers have also been provided.

The numerical examples illustrate the basic effect which accompanies contact of rough bodies, namely the inhomogeneity of deformation in a subsurface layer. This effect has already been discussed in Sect. 4.1 as the motivation for the micromechanical analysis of contact boundary layers in Chaps. 4, 5, and 6. In particular, it is solely due to the deformation inhomogeneities that plastic deformations can be induced in the vicinity of the surface, while the macroscopic deformations remain purely elastic. Note that significant shear deformation that is accumulated in the surface layer due to repeated ploughing by subsequent asperities in sliding contact. A relatively short sliding distance of three asperity spacings has only been simulated in the ploughing example of Sect. 6.3 because direct simulations of realistic sliding distances, which are normally at least 2–3 orders of magnitude greater, are computationally too expensive. The shakedown theory is a possible approach to treat the related problems of shakedown and ratchetting, cf. Johnson [57], although the finite deformation effects, e.g. the finite configuration changes and large slip contact, cannot be treated by the theory.

The real three-dimensional surface topography, measured by scanning profilometry, has been used to study the normal contact compliance of a sandblasted surface, and a very good agreement of the predicted normal contact compliance with the measured one has been obtained. Note that the associated computational effort was substantial, although the finite element discretization was not particularly fine with $55 \times 55 = 3025$ nodes on the contact surface. Nevertheless, in view of the continuous increase of available computing power, and also due to the continuous improvement of numerical algorithms, the present direct approach becomes more and more feasible. At the cost of high computational effort, direct simulations of contact interaction of rough surfaces offer several advantages with respect to the classical approach proposed by Greenwood and Williamson [34], for instance, the interactions between neighbouring asperities are naturally accounted for which is particularly important at high fractions of the real contact area.

The model developed in Sect. 6.4 is, however, not fully predictive because the actual distribution of plastic properties within the surface layer is not known, and thus a uniform yield stress, equal to that in the bulk material, could only be assumed in the simulations. The lack of knowledge of local material properties (e.g. yield stress) within surface layers constitutes thus a barrier for predictive micromechanical modelling of contact interactions.

In this context, techniques such as micro- and nano-indentation might prove useful for the determination of local material properties.

As discussed in Sect. 3.3, asperity flattening and real contact area are substantially affected by macroscopic plastic deformations in contact conditions typical for metal forming processes. According to the results of the present micromechanical studies, the analogous effects in the elasto-plastic asperity deformation regime seem to be less pronounced. For instance, quite surprisingly, the macroscopic ploughing friction coefficient appears to be practically insensitive to the macroscopic elastic in-plane strain. At the same time, a visible effect has been predicted in the case of the normal contact compliance of the sand-blasted surface, although the effect is not substantial in quantitative terms.

The averaging procedure introduced in Sect. 5.2, proved to be a useful tool for the quantitative analysis of inhomogeneous fields within contact boundary layers. For instance, the analysis of the average stresses within the boundary layer indicates that the residual stresses, which develop due to ploughing, compensate the macroscopic stress, so that the total stress in the vicinity of the ploughing asperity is not affected by the macroscopic stress far from the surface. This is suggested as an explanation, why the ploughing friction coefficient is not affected by the macroscopic in-plane strain.

In the classical micromechanics of heterogeneous materials, a representative volume element is usually subjected to one of three classical types of boundary conditions (periodic displacement fluctuation, linear displacement, uniform traction), cf. Fig. 2.1, which ensure that the Hill's lemma is satisfied. In the case of contact boundary layers, V-periodicity of the displacement correction, as introduced in Chap. 4, is *essential* for consistent modelling of deformations in boundary layers. For instance, a meaningful solution would not be obtained in the asperity ploughing example of Sect. 6.3 if linear displacement or uniform traction boundary condition was imposed – note the deformation pattern induced by ploughing shown in Fig. 6.5. Periodicity of surface roughness, which implies V-periodicity of displacement correction, is thus an important assumption of the present micromechanical framework. Its application for real surfaces, which normally are not periodic, requires that approximate models are developed in which periodicity of roughness is enforced by modifying the original roughness topography, cf. Sect. 6.4.

Let us, finally, mention possible extensions of the micromechanical framework developed above. Rough contact interactions studied in Chaps. 4 and 6 correspond to dry or boundary lubrication conditions. Recently, a micromechanical model of the thin-film *hydrodynamic lubrication* regime has been developed, in which asperity deformation is modelled by applying the present boundary layer approach, while the flow of lubricant separating the contact surfaces is described using the Reynolds equation; see Stupkiewicz and Marciniszyn [134] for preliminary results.

The present approach is also directly applicable for stationary *heat conduction* problems. In fact, heat transfer through a contact interface is associated with micro-inhomogeneities of temperature and heat flux at the scale of interacting asperities. Micromechanical analysis of boundary layers induced by a micro-inhomogeneous heat flux is thus a convenient framework for modelling of contact heat transfer, e.g. for the prediction of the effective heat transfer coefficient, cf. Stupkiewicz and Sadowski [142]. It would also be interesting to extend the present approach to non-stationary heat conduction problems and to problems of coupled thermomechanical contact of rough bodies. In both cases, the corresponding asymptotic analysis would probably require consideration of multiple spatial and temporal scales.

The analysis of contact boundary layers and the contact-related examples studied in this work are limited to the case of contact of a deformable body with a rigid obstacle. In general, the present framework could be directly applied also for two deformable rough bodies in contact. Two difficulties are, however, foreseen. Firstly, the finite element implementation of sliding of two deformable surfaces would require a specialized contact search algorithm that would correctly detect contacting points for arbitrary relative position of the two representative surface samples. This is merely a technical problem. The second difficulty is more fundamental. As periodicity is an essential element of the present approach, representative surface samples must be chosen in a way that periodicity of roughness holds for both surfaces. Although periodicity could be imposed artificially by modifying the roughness, as in Sect. 6.4, ensuring representativeness might lead to unacceptably large finite element models. Furthermore, if finite macroscopic deformations were allowed, then the assumption of periodicity, which must continuously hold during deformation, would impose a very restrictive constraint on macroscopic deformations of the two bodies. As a remedy, a two-scale approach could be applied, in essence similar to that adopted in Sect. 3.3.7, provided that the characteristic dimensions of asperities of the two surfaces are significantly different.

7

Introduction to Martensitic Microstructures

Abstract: This chapter is a brief introduction to shape memory alloys and to martensitic microstructures. Basic concepts, phenomena, and definitions concerning martensitic transformations in shape memory alloys are introduced as a basis for the developments presented in Chaps. 8 and 9. The crystallographic theory of martensite is discussed for internally twinned martensites and for internally faulted martensites, and some remarks are given on the microstructure and interfacial energy of the transition layer at the austenite–twinned martensite interface.

7.1 Martensitic Transformation in Shape Memory Alloys

Shape memory alloys are attracting more and more interest due to their unique behaviour and due to the related advanced applications, e.g. Bhattacharya [13], Morawiec [83], Otsuka and Wayman [94], Sun [145]. The shape memory effect, pseudoelastic behaviour, and other phenomena observed in these alloys are associated with the martensitic phase transformation and, more specifically, with the formation and evolution of the microstructures that accompany the phase transition.

The martensitic transformation is a first-order, diffusionless, solid-to-solid phase transformation. The parent phase, called the *austenite*, is stable at higher temperatures, and the product phase, called the *martensite*, is stable at lower temperatures. The transition temperature depends on the alloy, its composition, heat treatment, etc. The transformation can be induced either by changing the temperature (thermally-induced transformation), or by applying external loading (stress-induced transformation). The martensitic transformation is crystallographically reversible, however, as a dissipative process, it is thermodynamically irreversible.

The martensitic phase transformation is a symmetry lowering transformation, i.e. the crystallographic lattice of the austenite has higher symmetry than that of the martensite. As a result, several *variants* of martensite can appear. The martensite variants are crystallographically equivalent, but rotated with respect to each other. This is schematically illustrated in Fig. 7.1 for the simple example of the cubic-to-tetragonal transformation. The tetragonal unit cell of the martensite can be obtained by stretching the unit cell of the cubic austenite along one of the three axes parallel to the edges of the cubic cell. There are thus three crystallographically equivalent variants of martensite. Clearly, the transformation strain, so-called Bain strain, i.e. the strain associated with the deformation of the crystallographic lattice during transformation is different for each variant.

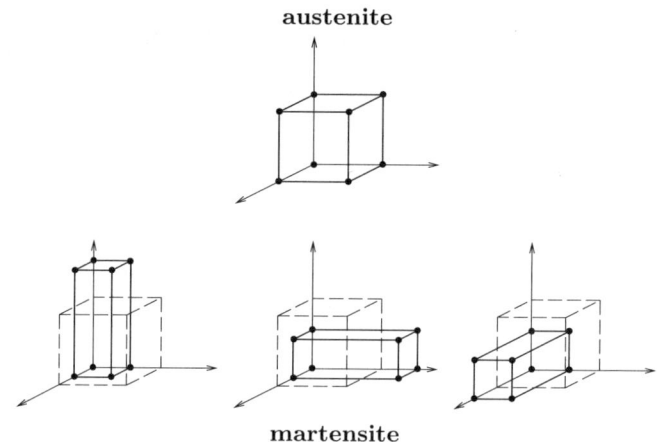

Fig. 7.1. Martensite variants in the cubic-to-tetragonal transformation.

The multiplicity of forms (variants) that the austenite can take upon transformation is, in fact, the source of the unique behaviour of shape memory alloys. The related phenomena are associated with the development and evolution of *martensitic microstructures*.

Consider first the transformation induced by decreasing the temperature at zero macroscopic stress. As there are no privileged directions, the austenite transforms into a mixture of all variants with essentially zero[1] macroscopic strain, as the transformation strains of particular variants cancel each other. The coexistence of austenite and martensite during transformation and the coexistence of different variants of martensite after transformation gives rise to martensitic microstructures. The basic microstructures, i.e. twins and austenite–martensite interfaces, are discussed in the next section.

Assume now that external loading is applied to the transformed specimen. Under stress, the martensite variants rearrange, so that the volume fraction of

---

[1] The volumetric component of transformation strain is usually very small in martensitic transformations in shape memory alloys.

some variants grows at the expense of the others. At the micro-scale, this occurs through migration of twin boundaries. Martensite variant rearrangement is associated with a macroscopic deformation which does not vanish after the stress is released, cf. Fig. 7.2. However, once the reverse transformation is induced by heating, the original shape is recovered in the austenitic state. Subsequent cooling, accompanied by austenite-to-martensite transformation at zero stress, produces again a mixture of all variants with zero macroscopic strain. The phenomenon corresponding to the thermomechanical loading cycle described above, and sketched in Fig. 7.2, is called the *shape-memory effect.*

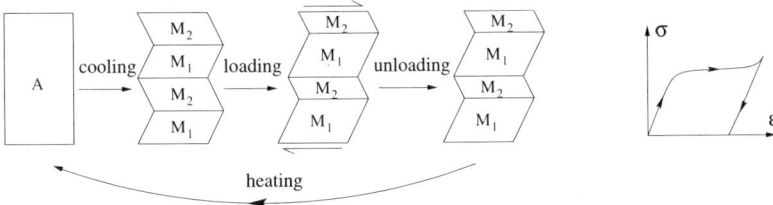

Fig. 7.2. Shape-memory effect.

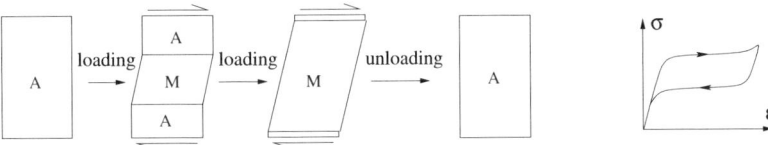

Fig. 7.3. Pseudoelasticity.

The martensitic transformation may also be induced by external loading at the temperature high enough for the austenite to be stable in the stress-free state. In that case, an *oriented* martensite appears, typically in the form of parallel plates, since only selected variants of martensite are suitably oriented with respect to the external stress. Once the whole volume is transformed, further loading results in the elastic loading of martensite, possibly accompanied by martensite variant rearrangement. Upon unloading, the reverse transformation proceeds with the typical hysteresis loop, as indicated in Fig. 7.3. The corresponding behaviour is referred to as *pseudoelasticity.* The stress-induced transformation can be modelled as an isothermal, i.e. purely mechanical, process under the assumption that the process is sufficiently slow. However, in general, the thermomechanical coupling may generate substantial effects.

The micromechanical modelling carried out in Chaps. 8 and 9 is concerned with the stress-induced transformations and with the related pseudoelastic effect. Accordingly, isothermal conditions are assumed, except in Sect. 8.3.4, where a macroscopically adiabatic process is studied.

7.2 Crystallographic Theory of Martensite

7.2.1 Geometrical Compatibility Condition

The *crystallographic theory of martensite* developed by Ball and James [8] is a rigorous mathematical theory extending the so-called *phenomenological theory of martensitic transformation* of Wechsler, Lieberman and Read [157] and Bowles and MacKenzie [17]. The theory, developed within the framework of nonlinear thermoelasticity, is widely accepted and highly successful in predicting the microstructures in martensitic transition. A detailed exposition of the theory with an extensive bibliography can be found in a recent book of Bhattacharya [13].

The crystallographic theory is based on the postulate that the microstructures are absolute minimizers of energy. Consequently, microstructure evolution and the related hysteresis are not addressed by the theory. Moreover, only stress-free (compatible) microstructures, whenever possible, are predicted by the theory. The final equations resulting from the theory are thus purely geometrical.

The martensitic transformation is displacive (i.e. diffusionless), and the related homogeneous deformation that transforms the lattice of the austenite to that of the martensite variant I is described by the *transformation stretch* tensor $\mathbf{U}_I$. For a specified pair of parent and product phases, the transformation stretch $\mathbf{U}_I$ is known, as it can be determined from the crystallographic lattice parameters of both phases. The undeformed, stress-free configuration of the austenite is adopted as a reference configuration, thus, in view of the assumption of zero stress, the deformation gradient within a transformed region occupied by single variant is $\mathbf{F}_I = \mathbf{R}\mathbf{U}_I$, where $\mathbf{R}$ is a proper rotation tensor, $\mathbf{R}\mathbf{R}^T = \mathbf{I}$ and $\det \mathbf{R} = 1$. Martensite variants are symmetry-related, thus, for each pair (I,J), there exists a rotation tensor $\mathbf{Q}_{IJ}$, belonging to the symmetry point group of austenite, which relates the respective transformation stretches, viz.

$$\mathbf{U}_J = \mathbf{Q}_{IJ}\mathbf{U}_I\mathbf{Q}_{IJ}^T. \tag{7.1}$$

As the deformation associated with the martensitic transformation is continuous (coherent transformation), the jump of deformation gradient across any interface, $\Delta\mathbf{F} = \mathbf{F}^- - \mathbf{F}^+$, must satisfy the following *geometrical compatibility condition*,

$$\Delta\mathbf{F} = \mathbf{c} \otimes \mathbf{n}, \tag{7.2}$$

which is the finite-strain counterpart to the compatibility condition (2.27). Here, $\mathbf{F}^+$ and $\mathbf{F}^-$ are the deformation gradients at both sides of the interface, $\mathbf{n}$ is the interface normal (in the reference configuration), and $\mathbf{c}$ is a vector.

Note that under the assumption of zero stress, the compatibility condition (7.2) enforces a purely geometrical constraint on transformation stretches of the martensite variants forming a microstructure. Clearly, the condition of mechanical equilibrium, i.e. the finite-strain counterpart to the compatibility condition (2.29), is trivially satisfied at zero stress.

The austenite–martensite interface is the basic microstructure in the analysis of progressive transformation when the austenite and the martensite coexist. This microstructure is discussed below. Other microstructures (e.g. wedges, complex twin microstructures) and thin-film microstructures (e.g. tents, tunnels) have been studied, for example, by Bhattacharya [11, 13], Bhattacharya and James [15], and Hane [39].

Consider first the case of the *exact* austenite–martensite interface, i.e. the interface formed by the austenite ($\mathbf{F}^+ = \mathbf{I}$) and the crystallographically perfect I-th variant of martensite ($\mathbf{F}^- = \mathbf{R}\mathbf{U}_I$). The compatibility condition (7.2) takes the form

$$\mathbf{R}\mathbf{U}_I - \mathbf{I} = \mathbf{b} \otimes \mathbf{m}, \tag{7.3}$$

to be solved for $\mathbf{m}$, $\mathbf{b}$, and $\mathbf{R}$ with $\mathbf{U}_I$ constituting the data. Here and in the following, $\mathbf{m}$ denotes the unit vector normal to the austenite–martensite interface, called the *habit plane*, and $\mathbf{b}$ is the so-called *shape strain* vector.

It can be shown that solution of (7.3) exists if and only if the transformation stretch $\mathbf{U}_I$ has one eigenvalue equal to one, one less than one, and one greater than one, see, for instance, Bhattacharya [13]. This condition is not satisfied by the vast majority of known shape memory alloys[2] and thus exact austenite–martensite interfaces are, in general, not possible. For this reason the austenite–martensite microstructures in shape memory alloys usually involve either twinning (*internally twinned martensites*) or stacking faults (*internally faulted martensites*), and these mechanisms allow geometrical compatibility between austenite and martensite at zero stress.

7.2.2 Internally Twinned Martensites

The internally twinned martensite appears in the form of a fine mixture of two variants. The microstructure of the corresponding austenite–martensite interface is depicted in Fig. 7.4. In addition to the compatibility condition at the austenite–martensite interface, the geometrical compatibility must hold along the *twinning plane*, i.e the martensite-martensite interface, with normal $\mathbf{l}$. This is expressed in the form of the *twinning equation*,

$$\mathbf{R}\mathbf{U}_I - \mathbf{U}_J = \mathbf{a} \otimes \mathbf{l}, \tag{7.4}$$

which, for a given pair (I, J) of martensite variants, is solved for unknown vectors $\mathbf{l}$ and $\mathbf{a}$, and rotation $\mathbf{R}$.

---

[2] Bhattacharya [13] reports two cases where, by careful manipulation of composition, the alloys have been found which satisfy condition (7.2) and form the exact austenite–martensite interface.

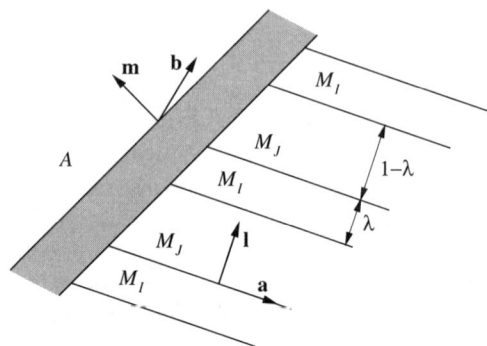

Fig. 7.4. Microstructure at the austenite-twinned martensite interface (A – austenite, M_I, M_J – martensite variants, in grey – a transition layer).

Assume now that solution of (7.4) exists, and consider a twin laminate formed by two martensite variants with λ and $(1 - \lambda)$ being the volume fractions of variants I and J, respectively. The austenite–martensite interface is now possible if the *average* deformation gradient of the twinned martensite satisfies the compatibility condition. This condition is called the *habit plane equation*, and can be written as

$$\hat{\mathbf{R}}[\lambda \mathbf{R} \mathbf{U}_I + (1 - \lambda) \mathbf{U}_J] - \mathbf{I} = \mathbf{b} \otimes \mathbf{m}. \tag{7.5}$$

The unknowns of the habit plane equation (7.5) are the habit plane normal $\mathbf{m}$, the shape strain vector $\mathbf{b}$, the twin fraction λ, and rotation $\hat{\mathbf{R}}$.

A detailed discussion of the conditions, under which solution of (7.4) and (7.5) exist, and the respective solution methods are provided by Bhattacharya [13] based on the results of Ball and James [8]. Examples and applications for specific alloys can be found, for example, in Bhattacharya [13] and Hane and Shield [41, 42, 43].

Note that the habit plane equation (7.5) is a *macroscopic* compatibility condition ensuring stress-free conditions everywhere except in a thin transition layer, depicted in grey in Fig. 7.4, in which the local incompatibilities must be accommodated by non-zero local elastic strains, see Sect. 7.3.

7.2.3 Internally Faulted Martensites

Although, as discussed above, the exact austenite–martensite interfaces are not possible in general, in some Cu-based alloys (e.g. CuAlNi, CuZnAl, CuZn, CuAlMn) coherent interfaces are observed between austenite and untwinned martensite. In these martensites, the compatibility at the austenite–martensite interface is obtained by macroscopic shear due to random stacking faults on the basal planes, cf. Otsuka et al. [95] and Chakravorty and Wayman [20], hence the name *internally faulted martensites*.

In the Cu-based shape memory alloys, usually more than one martensitic phase exists, and thus several austenite-to-martensite and also martensite-to-martensite transformations are possible, cf. Otsuka et al. [92, 95], Horikawa et al. [48], Dutkiewicz et al. [26], Šittner and Novak [153]. The internally faulted martensites appear in the transitions from the cubic austenite of DO_3 or B2 structure to the monoclinic martensite of 6M structure;[3] a typical example is the $\beta_1 \to \beta_1'$ transition in CuAlNi.

The specification of the crystallographic theory for the case of untwinned martensites is presented below based on the approach proposed by Hane [39]. However, following Stupkiewicz [131], an explicit distinction is made here between the shear resulting from the change of the crystallographic lattice and the additional shear due to random stacking faults.

The deformation which transforms the lattice of the cubic austenite to the crystallographically perfect lattice of the monoclinic 6M martensite can be decomposed into pure stretch and shear, the latter resulting from regular shuffling of the basal planes. Denote the corresponding deformation gradients of martensite variant I by $\mathbf{U}_I$ and $\mathbf{K}_I = \mathbf{I} + k_\theta \mathbf{s}_I \otimes \mathbf{n}_I$, respectively, where the stretch tensor $\mathbf{U}_I$ and the shear magnitude k_θ are known from the crystallography of transformation. The unit vectors $\mathbf{s}_I$, the direction of shear, and $\mathbf{n}_I$, the shear plane normal, are also known. The resulting deformation gradient,

$$\mathbf{F}_I = \mathbf{K}_I \mathbf{U}_I = (\mathbf{I} + k_\theta \mathbf{s}_I \otimes \mathbf{n}_I)\mathbf{U}_I, \tag{7.6}$$

does not, in general, satisfy the geometrical compatibility condition (7.3), namely $\mathbf{R}\mathbf{F}_I - \mathbf{I} \neq \mathbf{b} \otimes \mathbf{m}$.

As already mentioned, the compatibility at the austenite–martensite interface is provided by random stacking faults on the basal planes. The associated deformation is a shear,

$$\mathbf{K}_I^{\mathrm{sf}} = \mathbf{I} + k_{\mathrm{sf}} \mathbf{s}_I \otimes \mathbf{n}_I, \tag{7.7}$$

where k_{sf} is an unknown shear magnitude due to stacking faults. Importantly, the shear system is the same as the one in $\mathbf{K}_I$. The mechanism of stacking faulting is through the *sequence faults*, i.e by the perturbation of the ideal shuffling sequence responsible for the shear $\mathbf{K}_I$, cf. Andrade et al. [4], see also Sect. 9.2.1.

The geometrical compatibility condition is now enforced on the total deformation gradient $\hat{\mathbf{F}}_I = \mathbf{K}_I^{\mathrm{sf}} \mathbf{F}_I$, namely

$$\mathbf{R}[\mathbf{I} + (k_\theta + k_{\mathrm{sf}})\mathbf{s}_I \otimes \mathbf{n}_I]\mathbf{U}_I - \mathbf{I} = \mathbf{b} \otimes \mathbf{m}, \tag{7.8}$$

and $\hat{\mathbf{F}}_I$ is given by

---

[3] The 6M structure can alternatively be represented by the M18R or M9R structure, depending on the parent phase structure, but the 6M unit cell corresponds to the actual transformation mechanism and more accurately reflects the symmetry of the product phase, cf. Otsuka et al. [91]. The transformation mechanism and different unit cells are discussed in more detail in Sect. 9.2.1.

$$\hat{\mathbf{F}}_I = \mathbf{K}_I^{\text{sf}}\mathbf{K}_I\mathbf{U}_I = [\mathbf{I} + (k_\theta + k_{\text{sf}})\mathbf{s}_I \otimes \mathbf{n}_I]\mathbf{U}_I. \tag{7.9}$$

The unknowns in the habit plane equation (7.8) are the habit plane normal
$\mathbf{m}$, the shape strain vector $\mathbf{b}$, the shear magnitude k_{sf}, and a rotation $\mathbf{R}$. The
solution of (7.8) can be obtained by applying the theory of Ball and James [8],
for more details refer to Hane [39] and Stupkiewicz [131].

Remark 7.1. In the present setting, the stacking faults are assumed to appear
freely if this is required by the compatibility of the total deformation gradient
$\hat{\mathbf{F}}_I^{\text{t}}$, and thus the shear magnitude k_{sf} can take arbitrary values at no energetic
expense. This is in contrast to the analysis of Chap. 9 where the stacking faults
are assumed to increase the free energy of the martensite.

7.2.4 Geometrically Linear Theory

Although the kinematically exact crystallographic theory of martensite is readily available, the geometrically linear one is briefly introduced below. Being
an approximation, it is, however, consistent with the small deformation framework of the micromechanical analysis carried out in Chaps. 8 and 9.

Neglecting the difference between the gradients in the reference and current configurations, the small strain tensor ε corresponding to a deformation
gradient $\mathbf{F}$ is derived from the displacement gradient $\nabla \mathbf{u} = \mathbf{F} - \mathbf{I}$ according
to

$$\varepsilon = \frac{1}{2}[\nabla \mathbf{u} + (\nabla \mathbf{u})^{\text{T}}] = \frac{1}{2}(\mathbf{F} + \mathbf{F}^{\text{T}}) - \mathbf{I}. \tag{7.10}$$

Thus, for example, the *transformation strain* ε_I^{t} corresponding to the transformation stretch $\mathbf{U}_I$ is given by $\varepsilon_I^{\text{t}} = \mathbf{U}_I - \mathbf{I}$. As in the case of the transformation stretches, the transformation strains of any two variants of martensite
are mutually rotated, so that $\varepsilon_J^{\text{t}} = \mathbf{Q}_{IJ}\varepsilon_I^{\text{t}}\mathbf{Q}_{IJ}^T$, cf. (7.1).

Consider first the internally twinned martensites. The twinning equation,
i.e. the small strain counterpart to (7.4), reads

$$\varepsilon_I^{\text{t}} - \varepsilon_J^{\text{t}} = \frac{1}{2}(\mathbf{a} \otimes \mathbf{l} + \mathbf{l} \otimes \mathbf{a}), \tag{7.11}$$

and the habit plane equation (7.5) becomes

$$\lambda\varepsilon_I^{\text{t}} + (1 - \lambda)\varepsilon_J^{\text{t}} = \frac{1}{2}(\mathbf{b} \otimes \mathbf{m} + \mathbf{m} \otimes \mathbf{b}), \tag{7.12}$$

where ε_I^{t} and ε_J^{t} are the transformation strains of martensite variants I and
J, respectively, and the strain in the austenite is equal to zero. As in the
geometrically nonlinear theory the unknowns are $\mathbf{l}$, $\mathbf{a}$, $\mathbf{m}$, $\mathbf{b}$, and λ.

In the case of internally faulted martensites, the total transformation strain
$\hat{\varepsilon}_I^{\text{t}}$ can be expressed as a sum of the transformation strain ε_I^{t} of an un-faulted
martensite and the additional shear due to stacking faults,

$$\hat{\varepsilon}_I^{\mathrm{t}} = \varepsilon_I^{\mathrm{t}} + k_{\mathrm{sf}} \frac{1}{2} (\mathbf{s}_I \otimes \mathbf{n}_I + \mathbf{n}_I \otimes \mathbf{s}_I). \tag{7.13}$$

The habit plane equation, expressing the geometrical compatibility of the austenite and the faulted martensite, takes now the form

$$\varepsilon_I^{\mathrm{t}} + k_{\mathrm{sf}} \frac{1}{2} (\mathbf{s}_I \otimes \mathbf{n}_I + \mathbf{n}_I \otimes \mathbf{s}_I) = \frac{1}{2} (\mathbf{b} \otimes \mathbf{m} + \mathbf{m} \otimes \mathbf{b}), \tag{7.14}$$

with unknown $\mathbf{m}$, $\mathbf{b}$, and k_{sf}.

The method of solution of (7.11) and (7.12) can be found in Bhattacharya [12], and its specification for (7.14), i.e. for the case of internally faulted martensites, is provided in Stupkiewicz [131]. The predictions of the geometrically linear theory are, in general, close to those of the kinematically exact one. There are, however, some qualitative differences due to rotations which are neglected in the linear theory, cf. Bhattacharya [12].

7.3 Transition Layer at the Austenite–Twinned Martensite Interface

Let us have a closer look at the austenite–martensite interface in internally twinned martensites. Macroscopically, this well defined interface (the habit plane) separates the austenite and the martensite, and its orientation and other microstructural features can be predicted using the crystallographic theory outlined above. However, at the micro-scale, the local incompatibility of the undeformed austenite with the martensite variants taken separately must be accommodated by nonzero elastic strains. These micro-strains are confined to a relatively thin transition (boundary) layer depicted schematically in grey in Fig. 7.4. The corresponding elastic energy, interpreted as the interfacial energy at a higher scale of observation, and the related effects are briefly discussed below.

It is well recognized that relationships between characteristic microstructural dimensions at different scales are determined by size-dependent contributions of different kinds of bulk and interfacial energy to the total free energy. For instance, the twin spacing results from the interplay of the elastic micro-strain energy at the austenite–martensite interface and the interfacial energy of twin boundaries, see for example Khachaturyan [60], Kohn and Müller [64], Roytburd [120], Petryk et al. [104].

The tendency of the material to minimize the total energy may lead to twin branching (i.e. to the refinement of twin spacing in the vicinity of the austenite–martensite interface: the elastic micro-strain energy decreases as the twin spacing decreases) as well as to the formation of complex microstructures of the transition layer itself. This has been studied theoretically, cf. Ball and James [8], James et al. [53], Kohn and Müller [64], and also observed in experiments, cf. Hÿtch et al. [50], Liu and Dunne [80], see also Figs. 7.2, 7.4 and

7.6 in Bhattacharya [13]. In general, these experimental observations suggest that the interface is not sharp at the micro-scale.

From dimensional analysis it follows that the density γ_e of interfacial energy of elastic micro-strains, per unit nominal area of the macroscopic austenite–martensite interface, is proportional to the twin spacing h. Thus a size-independent parameter Γ_e can be introduced,

$$\Gamma_e = \frac{\gamma_e}{h}, \qquad (7.15)$$

which is a characteristic quantity of the microstructure of the transition layer, i.e. of the shape of the interface at the micro-scale.

For an *a priori* assumed microstructure of the transition layer, the elastic micro-strain energy can be computed by solving the problem of theory of elasticity with the eigenstrains being constant and given within each phase. For example, Sridhar et al. [129] presented an analytical solution for a simple microstructure (with the phases occupying rectangular domains) in the case of linear elasticity, the whole system being elastically isotropic and homogeneous. A simple estimate of the elastic micro-strain energy, calibrated using the results of Sridhar et al. [129], has been provided by Roytburd [120].

A more general micromechanical approach has recently been developed by Maciejewski et al. [82], and the main results of this work are briefly outlined below. A class of zigzag shapes of the austenite–martensite interface has been considered with angle θ defining the deviation of the zigzag-shaped interface from the planar one, cf. Fig. 7.5. Periodic microstructure has then been assumed, and a two-dimensional boundary value problem has been formulated for the unit cell – the plane of analysis being defined by unit vectors **m** and **l** normal to the habit plane and twinning plane, respectively. Twin branching has not been directly accounted for, thus the twin spacing h refers to the finest microstructure in the vicinity of the austenite–martensite interface. The finite deformation setting has been adopted, and the elastic anisotropy of the phases has been fully accounted for. Finite element solution of the boundary value problem for the unit cell provides the elastic energy as well as the local strains and stresses within the transition layer. As expected, in macroscopically stress-free conditions, the latter are concentrated in the vicinity of the interface and vanish far from the interface. The interfacial energy parameter Γ_e is obtained from the total elastic energy of micro-strains.

The above micromechanical scheme has been applied for the cubic-to-orthorhombic $(\beta_1 \rightarrow \gamma_1')$ transformation in CuAlNi alloy. According to the crystallographic theory, in this transformation there are 96 distinct microstructures of the austenite–martensite interface which can be divided into four groups, each consisting of 24 crystallographically equivalent microstructures, see Appendix B.

The results of finite element computations are summarized in Fig. 7.6, where the computed values of parameter Γ_e are shown as a function of the wedge angle θ for four microstructures denoted by M1,...M4 (one from each

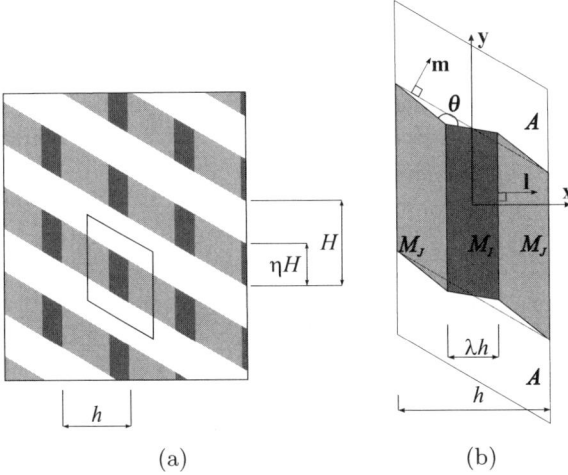

(a)   (b)

Fig. 7.5. Schematic view of the analyzed microstructure: (a) austenite–martensite laminate, (b) periodic unit cell (reprinted from [82], Copyright 2005, with permission from IFTR PAS).

group of 24 crystallographically equivalent microstructures). It is seen that the shape of interface strongly affects the micro-strain energy. In particular, deviation from the planar interface may significantly reduce the energy (the vertical dashed line at $\theta = 180°$ in Fig. 7.6 indicates the planar interface).

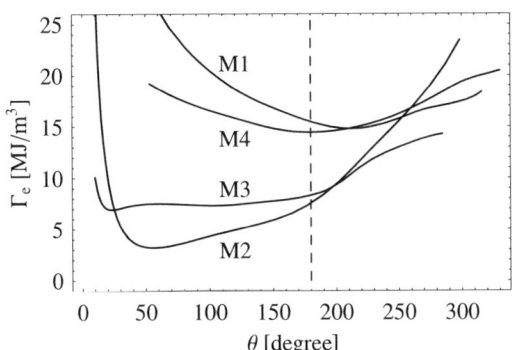

Fig. 7.6. Elastic micro-strain energy parameter Γ_e as a function of the wedge angle θ of a zigzag-shaped austenite–twinned martensite interface (reprinted from [82], Copyright 2005, with permission from IFTR PAS).

The shapes of the minimum energy interfaces are shown in Fig. 7.7 for the four microstructures. Note, however, that the morphology of the interface may also be affected by other factors, for instance by the interfacial energy of direct austenite–martensite boundaries. This energy increases with increasing area of the interface at the micro-scale, thus in a sense it penalizes deviation from the planar interface. The relative importance of this interfacial energy and of the elastic micro-strain energy is size dependent: for a given shape of the interface the latter scales linearly with twin spacing h while the former is constant, cf. Maciejewski et al. [82].

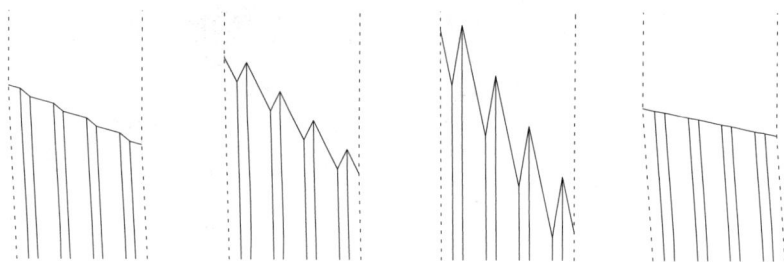

Microstructure M1 Microstructure M2 Microstructure M3 Microstructure M4

Fig. 7.7. Austenite–twinned martensite interface shapes corresponding to the minimum values of elastic micro-strain energy parameter Γ_e, cf. Fig. 7.6.

8

Evolving Laminates in Shape Memory Alloys

Abstract: Micromechanical model of a single crystal undergoing stress-induced martensitic transformation is developed assuming that the transformation proceeds by the nucleation and growth of parallel martensitic plates. The microstructure accompanying the transformation is thus that of an evolving rank-one laminate or of an evolving rank-two laminate if detwinning is additionally accounted for. Propagation of the phase transformation front is assumed to be governed by a rate-independent criterion formulated in terms of the thermodynamic driving force on the phase transformation front. Macroscopic constitutive rate-equations are derived for the case of an evolving rank-one laminate. Several examples of pseudoelastic response of Cu-based shape memory alloys illustrate the approach.

8.1 Laminated Microstructures

As indicated in the previous chapter, the coexistence of stress-free parent and product phases during a coherent phase transformation is only possible if the two phases are geometrically compatible. The condition of geometrical compatibility under zero stress is, in fact, a fundamental relation of the crystallographic theory of martensite. One of the results of that theory is that the orientation of the austenite–martensite interface, if such an interface exists, is fully determined by the transformation strains of martensite variants.

The martensitic transformations are thus accompanied by microstructures, the simplest being that associated with the interface between the austenite and the twinned or untwinned martensite. These basic microstructures are also involved in more complex microstructures, such as wedges, crossing plates, etc., cf. Bhattacharya [11], Shield [126].

In stress-induced transformations, the assumption of zero stress is naturally not satisfied. However, in view of the elastic strains being small compared to the transformation strain, the predictions of the crystallographic theory are often regarded sufficiently accurate, and are thus applied also for the stress-induced transformations. For instance, Chu [22] and Ball et al. [7] developed the so-called *constrained theory*, applicable for stress-induced transformations, in which the elastic strains are neglected. The theory employs the finite deformation kinematics, and is based on minimization of the total free energy of a stressed specimen, which includes the potential energy of the loading device. The theory is consistent with the crystallographic theory, however, it is based on a simplifying assumption of stress homogeneity. Moreover, it does not account for hysteresis.

A different approach is adopted in the present work. The transformation is assumed to proceed through the evolution of laminated microstructures. Such microstructures are commonly observed in experiments, e.g. Otsuka et al. [95], Horikawa et al. [48], Jiang and Xu [54], Huo and Müller [49], Shield [126]. Laminated microstructures are also very convenient for micromechanical modelling because analytical micro-macro transition relations are available, see Sect. 2.6. While there are some similarities of the present approach to that of Khachaturyan [59, 60] and Roytburd [117, 118, 123], there are some fundamental differences. First of all, the microstructure evolution is determined here from the threshold condition imposed on the local thermodynamic driving force at the phase transformation front, rather than from the condition of phase equilibrium. Secondly, the elastic anisotropy of the phases with the related elastic mismatch between austenite and martensite are fully accounted for, and a complete computational scheme for the evolution of laminated microstructures is specified. In Chap. 9, an additional aspect is also discussed, namely the energy of stacking faults in internally faulted martensites.

In the simplest case, the laminated microstructure consists of parallel plates of martensite that nucleate and grow within the austenite matrix. The related evolving rank-one laminate is schematically illustrated in Fig. 8.1. Because of the applied external loading, some directions and some variants of martensite are favored. This is in contrast to the thermally-induced transformations, where more complex, but un-oriented, microstructures develop providing zero macroscopic strain.

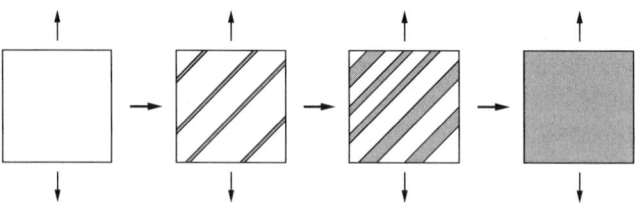

Fig. 8.1. Stress-induced transformation proceeding by formation and growth of parallel martensitic plates (evolving rank-one laminate).

In some situations, higher-rank laminates are also observed. For example, if the transformation proceeds through the formation and growth of one family of martensitic plates, as indicated in Fig. 8.1, but the martensitic plates are internally twinned, then the microstructure is a rank-two laminate, as illustrated in Fig. 8.2(a). Evolving rank-two laminates are studied in Sect. 8.4, in the context of martensite variant rearrangement (detwinning) during progressive transformation.

Higher-rank laminates are also observed, for instance, when secondary plates develop, cf. Fig. 8.2(b), or when instability of the macroscopically homogeneous transformation occurs. The related effects are not discussed here, some preliminary results can be found in Petryk and Stupkiewicz [103].

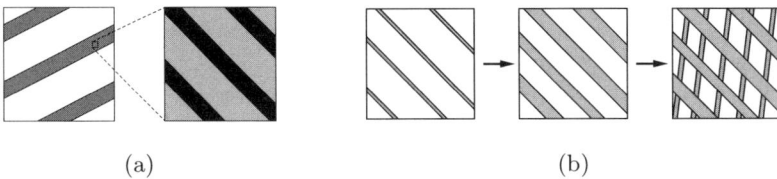

(a)   (b)

Fig. 8.2. Examples of evolving higher-rank laminates: (a) twinned martensitic plates, (b) secondary plates.

The present chapter is organized as follows. In Sect. 8.2, the small-strain constitutive framework and the transformation criterion are specified, and the macroscopic rate-equations are derived for the case of an evolving rank-one laminated microstructure. This section essentially follows the micromechanical modelling of Stupkiewicz and Petryk [140], and also that of Petryk [102] who developed the general finite deformation framework. The micromechanical model is next applied to simulate the pseudoelastic behaviour of CuZnAl and CuAlNi shape memory alloys. Evolution of rank-one laminates is studied in Sect. 8.3, and the effect of detwinning is analyzed in Sect. 8.4 by considering the evolution of a rank-two laminated microstructure (after Stupkiewicz and Petryk [141]).

8.2 Micromechanical Model of Pseudoelasticity

8.2.1 Constitutive Framework

There are two basic approaches to describe the constitutive behaviour of a single crystal undergoing martensitic phase transformation. In the first approach, the phases are explicitly considered to be different states of the same material. Consequently, a single free energy function is defined with local minima corresponding to austenite and to martensite variants, cf. Fig. 8.3(a). The phase of a material point is then determined by proximity to the nearest local minimum (so-called energy well). In this approach, the microstructure is typically found by minimizing the total free energy of an element, and, as the free energy function is not convex, relaxation techniques are usually applied, e.g. Bhattacharya and Dolzmann [14], Kohn [63], Smyshlyaev and Willis [128]. In a general case, it is not easy to formulate a free energy function corresponding to an arbitrary transformation and satisfying the elastic symmetries consistent with crystal symmetries in the wells, cf. Vedantam and Abeyaratne [151]. As a result, the applicability of that approach for real materials is restricted.

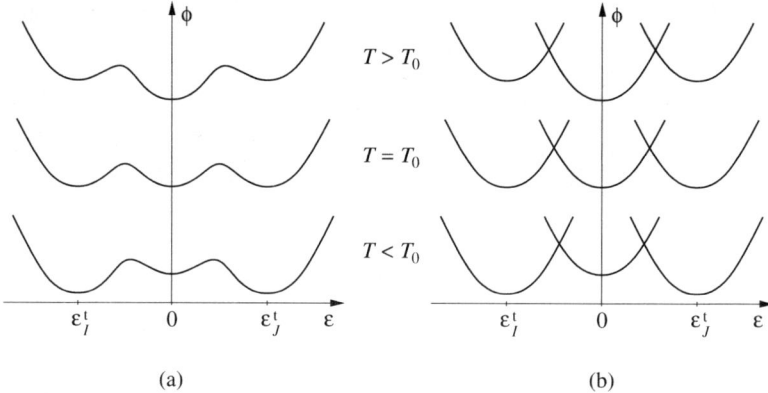

(a)   (b)

Fig. 8.3. Constitutive modelling of materials undergoing phase transformation: (a) single multi-well free energy function; (b) free energy functions prescribed separately for each phase (to be complemented by a transformation criterion).

In the second approach, which is adopted in the present considerations, a micro-structured material undergoing phase transformation is treated as a mixture of separate phases. The austenite and each variant of martensite are assumed to be elastic materials with usual convex free energy functions, each having a distinct stress-free configuration, cf. Fig. 8.3(b). The phase transformation is accounted for by enforcing transformation criteria at the interfaces. In this approach, the symmetry relations between the phases are easily introduced within the framework of anisotropic elasticity with eigenstrain. The small-strain constitutive relationships are provided below, however, an extension to finite deformation kinematics is also possible, cf. Maciejewski et al. [82], Stupkiewicz and Petryk [139].

Within the small deformation framework, the stress-free configuration of the martensite variant I is described by its transformation strain ε_I^t, measured with respect to the stress-free configuration of austenite, cf. Sect. 7.2.4. Consider thus the Helmholtz free energy function of the martensite variant I, as in the classical linear thermoelasticity, cf. Raniecki [110],

$$\phi_I(T, \varepsilon) = \phi_I^0(T) + \frac{1}{2}\left(\varepsilon - \varepsilon_I^t\right) \cdot \mathbf{L}_I\left(\varepsilon - \varepsilon_I^t\right), \tag{8.1}$$

where T denotes the temperature and

$$\phi_I^0(T) = \rho c[T - T_0 - T\log(T/T_0)] + u_m^0 - Ts_m^0, \quad \varepsilon_I^t = \varepsilon_I^{t0} + \alpha_I(T - T_0). \tag{8.2}$$

Here, ϕ_I^0 is the temperature-dependent free energy density in the stress-free state, ρ is the density, c is the specific heat, u_m^0 is the internal energy density in the reference state (i.e. at $T = T_0$ and in stress-free conditions), s_m^0 is the entropy in the reference state, and all densities refer to unit volume. Clearly, the internal energy in the reference state u_m^0 and the entropy in the reference state s_m^0 are identical for all martensite variants. In general, due to thermal expansion, the eigenstrain ε_I^t, depends on temperature, ε_I^{t0} is the eigenstrain at the reference temperature, and α_I is the linear thermal expansion tensor. Finally, $\mathbf{L}_I$ is the fourth-order elasticity tensor.

The free energy density of the austenite has the form analogous to (8.1)–(8.2) with the material constants u_a^0, s_a^0, $\mathbf{L}_a$, and α_a replacing u_m^0, s_m^0, $\mathbf{L}_I$, and α_I, respectively, and the specific heat of the austenite is assumed to be equal to that of the martensite. As all the strains are referred to the stress-free configuration of austenite at the reference temperature T_0, the eigenstrain of the austenite at $T = T_0$ is equal to zero.

In the following, the isothermal pseudoelastic response is mostly considered, thus a simplified form of the free energy is of interest. The *isothermal free energy* density of the austenite, consistent with the general form specified by (8.1)–(8.2), can be written in the form

$$\phi_a = \phi_0 + \frac{1}{2}\varepsilon \cdot \mathbf{L}_a \varepsilon, \tag{8.3}$$

and that of the martensite variant I in the form

$$\phi_I = \phi_0 + \Delta^{am}\phi_0 + \frac{1}{2}\left(\varepsilon - \varepsilon_I^t\right) \cdot \mathbf{L}_I\left(\varepsilon - \varepsilon_I^t\right), \tag{8.4}$$

where ϕ_0 and $\phi_0 + \Delta^{am}\phi_0$ are the temperature-dependent free energy densities of, respectively, austenite and martensite in stress-free states. The difference, denoted by $\Delta^{am}\phi_0$, is often called the *chemical energy*, and, in view of $(8.2)_1$, it can be expressed in the form

$$\Delta^{am}\phi_0 = -(u_a^0 - u_m^0) + T(s_a^0 - s_m^0), \tag{8.5}$$

where $u_a^0 - u_m^0 > 0$ and $s_a^0 - s_m^0 > 0$, cf. Raniecki [110]. It is seen that, in accord with experimental observations, the chemical energy $\Delta^{am}\phi_0$ is a linear function of temperature. The free energy densities (8.3) and (8.4) are consistent with the constitutive equations of linear isothermal anisotropic elasticity with eigenstrain, cf. (2.32), and the fourth-order elasticity tensors $\mathbf{L}_a$ and $\mathbf{L}_I$ possess the usual symmetries and are positive definite.

In view of the symmetry relations between the martensite variants, the transformation strains ε_I^t and ε_J^t of any two variants I and J are mutually rotated, so that $\varepsilon_J^t = \mathbf{Q}_{IJ}\varepsilon_I^t\mathbf{Q}_{IJ}^T$, see also (7.1). Also the elasticity tensors $\mathbf{L}_I$ and $\mathbf{L}_J$ are mutually rotated, so that, in Cartesian coordinates, we have $(L_J)_{ijkl} = (Q_{IJ})_{ip}(Q_{IJ})_{jq}(Q_{IJ})_{kr}(Q_{IJ})_{ls}(L_J)_{pqrs}$.

8.2.2 Phase Transformation Criterion

The martensitic transformation is crystallographically reversible in the sense that an identical crystalline lattice is recovered after the complete forward–reverse transformation cycle. However, the martensitic transformation is not thermodynamically reversible because it is a dissipative process. Indeed, hysteresis loops are observed in the case of both thermally- and stress-induced transformations.

The intrinsic dissipation rate due to the transformation, per unit area of the transformation front, is given by

$$\dot{\mathcal{D}}^t = fv_n \geq 0, \tag{8.6}$$

where f is the local *thermodynamic driving force* and $v_n \geq 0$ is the normal speed of propagation. In the quasi-static case,[1] the thermodynamic driving force is defined by, cf. Eshelby [28], Rice [114], Raniecki [110],

$$f = \boldsymbol{\sigma} \cdot \Delta\boldsymbol{\varepsilon} - \Delta\phi, \tag{8.7}$$

where ϕ is the Helmholtz free energy density, and $\Delta(\cdot) = (\cdot)^- - (\cdot)^+$ denotes the forward jump with respect to time of a local variable on the transformation front moving from '$-$' to '$+$' side, cf. Petryk [102]. In view of the property $\Delta\boldsymbol{\sigma} \cdot \Delta\boldsymbol{\varepsilon} = 0$, cf. (2.31), the stress $\boldsymbol{\sigma}$ in the formula (8.7) can be taken from any side of the front.

The thermodynamic driving force f may be used to formulate the criterion of propagation of the phase transformation front. As the dissipation of energy does not approach zero with decreasing rate of change of external conditions, it is commonly accepted that a barrier exists for f, so that the martensitic transformation can only proceed when f reaches some threshold value f_c. For instance, Abeyaratne and Knowles [1, 2] assumed a kinetic relation, i.e. a rate-dependent rule in which the propagation speed v_n is a function of the thermodynamic driving force f.

In this work, following Petryk [102] and Stupkiewicz and Petryk [140], a *rate-independent* criterion is adopted. The time-dependent effects are thus disregarded based on the assumption that the external displacements or loads change sufficiently slowly. This assumption is also consistent with the

---

[1] The thermodynamic driving force in the dynamic case has been derived by Abeyaratne and Knowles [1] and Raniecki and Tanaka [113]. In the present work, the inertia effects are neglected.

assumption that the process of stress-induced phase transformation is isothermal. Assume thus that the phase transformation front propagates if the
thermodynamic driving force attains a threshold value f_c, cf. Rice [114], Raniecki [110],

$$f - f_c \leq 0, \qquad v_n \geq 0, \qquad (f - f_c)v_n = 0. \tag{8.8}$$

The critical driving force f_c is, in general, state-dependent, furthermore, the
non-negativeness of the intrinsic dissipation rate requires that f_c is non-
negative, $f_c \geq 0$.

Transformation criterion (8.8) applies to both austenite-to-martensite and
martensite-to-austenite transitions with the proper meaning of the jumps in
the definition (8.7) of the driving force f. In general, the critical driving force
f_c may be different for the forward and for the reverse transformation.

If the critical driving force is assumed to be a material constant, $f_c =$
const., then, in the absence of dissipation due to reorientation of martensite
variants, the intrinsic dissipation $\bar{\mathcal{D}}^t$ due to any transformation in a finite
volume V^t is, cf. Stupkiewicz and Petryk [140],

$$\bar{\mathcal{D}}^t = V^t f_c. \tag{8.9}$$

If, additionally, the critical driving forces for forward and reverse transformations are identical, then the area of a hysteresis loop on an isothermal
stress–strain diagram is always equal to $2f_c$ and constant irrespective of the
loading program (e.g. tension or compression, loading direction, etc.), provided
that the whole volume transforms during forward and during reverse transformation. This conclusion is consistent with several experimental observations.
For example, the hysteresis size is found to be independent of temperature
(e.g. Huo and Müller [49], Shield [126]) and approximately equal in tension
and compression (e.g. Orgeas and Favier [90], Thamburaja and Anand [149]).
The criterion (8.8) is, however, not consistent with the special character of
the *internal* hysteresis loops observed by Müller and Xu [85] and Huo and
Müller [49].

Remark 8.1. If $f_c = 0$, then the transformation criterion (8.8) reduces to the
well-known condition of phase equilibrium, cf. Roytburd [117, 119], James [52],
Gurtin [37], Raniecki [110].

8.2.3 Macroscopic Rate-Equations for Laminates

In this section, the micro-macro transition relations are derived for a representative volume element of a material undergoing martensitic phase transformation. It is assumed that the microstructure accompanying the transformation
is that of an evolving rank-one laminate, cf. Fig. 8.1. The product phase is
thus assumed to appear in the form of parallel plates which nucleate and grow
within the parent phase. It is also assumed that no microstructural changes
appear within the layers.

The analysis below is based on that carried out in Petryk [102] and in Stupkiewicz and Petryk [140]. However, the present exposition is restricted to a simplified case, as the evolution of rank-one laminated microstructures is only considered here. More general forms of some formulae that are derived below can be found in Petryk [102] and in Stupkiewicz and Petryk [140]. The analysis of Petryk [102] employs the finite deformation framework, while here, and in Stupkiewicz and Petryk [140], a restriction is made to the infinitesimal strain format.

Consider thus a rank-one laminated microstructure within a material (single crystal) with $\eta^- = \eta$ and $\eta^+ = (1 - \eta)$ denoting the volume fractions of two homogeneous '−' and '+' phases. The product phase, denoted by '−', is identified with the martensite when the austenite-to-martensite transition is considered, and with the austenite in the case of reverse transformation. In agreement with (8.3)–(8.4), the constitutive behaviour of the phases is governed by the free energy density in the form

$$\phi^\pm = \phi_0^\pm + \frac{1}{2}\left(\varepsilon^\pm - \varepsilon^{t\pm}\right) \cdot \mathbf{L}^\pm\left(\varepsilon^\pm - \varepsilon^{t\pm}\right), \qquad (8.10)$$

so that $\boldsymbol{\sigma}^\pm = \mathbf{L}^\pm(\varepsilon^\pm - \varepsilon^{t\pm})$, cf. (2.32) and (2.34). The local transformation strains $\varepsilon^{t\pm}$ are assumed constant and known.

For a *fixed* microstructure, all the quantities, both macroscopic and microscopic, can be determined using the micro-macro transition relations for the rank-one laminated microstructure at hand, cf. Sect. 2.6 and Appendix A. However, this work is focused on the progressive transformation and on the related *evolution* of the microstructure. Therefore, the macroscopic rate-quantities are studied below.

In the process of martensitic transformation, the mechanical equilibrium can only be assumed at a discrete set of states which are separated by dynamic formation of product-phase particles (platelets). However, in the present analysis, the microstructure evolution is assumed to be continuous and quasistatic. This idealization is based on the assumptions that the overall strains vary slowly and that the thickness of each newly formed platelet is sufficiently small. The evolution can thus be described in terms of *forward rates*, i.e. right-hand derivatives, denoted by a superimposed dot.

As discussed in Sect. 2.6, in a laminate, strains and stresses are uniform within the layers. Thus the conditions are identical at all interfaces separating the layers, and the interface normal $\mathbf{n}$ can be assumed constant[2] during the

---

[2] Within the present modelling, the microstructure is assumed to be fine, actually infinitely fine in the limit. The gradients of macroscopic quantities as well as the boundary effects are thus disregarded. As a consequence, the variation of the interface normal $\mathbf{n}$ during transformation is also ruled out, because the related rotation of the interface would induce reverse transformation on a part of each interface, in contradiction with the homogeneity of stresses and strains within the layers, and with the homogeneity of the thermodynamic driving force along the interfaces.

transformation. Furthermore, in the present case of a two-phase laminate, the macroscopic strain and stress are simply given by

$$\mathbf{E} = \{\varepsilon\} = \eta\varepsilon^- + (1-\eta)\varepsilon^+, \qquad \mathbf{\Sigma} = \{\sigma\} = \eta\sigma^- + (1-\eta)\sigma^+, \qquad (8.11)$$

as a specification of the averaging operation (2.8).

An important consequence of evolution of the microstructure is that the rates of the macroscopic strain and stress are distinct from the respective averaged local rate-variables. Indeed, by differentiating (8.11), the rates of macroscopic stress and strain are obtained in the form

$$\boxed{\dot{\mathbf{E}} = \{\dot\varepsilon\} + \dot\eta\Delta\varepsilon, \qquad \dot{\mathbf{\Sigma}} = \{\dot\sigma\} + \dot\eta\Delta\sigma,} \qquad (8.12)$$

see Petryk [102] for the derivation of a more general *extended transport theorem*.

The local stresses and strains in the layers satisfy the compatibility conditions (2.27) and (2.29). As the interface normal $\mathbf{n}$ is fixed in space and time, similar compatibility conditions hold also for the rates of local variables, namely

$$\Delta\dot\varepsilon = \frac{1}{2}(\mathbf{d}\otimes\mathbf{n} + \mathbf{n}\otimes\mathbf{d}), \qquad \Delta\dot\sigma\mathbf{n} = \mathbf{0}. \qquad (8.13)$$

Using the interior–exterior decomposition, the above compatibility conditions can be rewritten in the form

$$\dot{\mathbf{E}}_P = \{\dot\varepsilon_P\} = \dot\varepsilon_P^+ = \dot\varepsilon_P^-, \qquad \dot{\mathbf{\Sigma}}_A = \{\dot\sigma_A\} = \dot\sigma_A^+ = \dot\sigma_A^-, \qquad (8.14)$$

see also (2.41). Note that, in view of compatibility conditions (2.30), equations (8.14) are not in contradiction to (8.12).

As the local transformation strains $\varepsilon^{t\pm}$ are constant, the rates of local strains and stresses are related by $\dot\varepsilon^\pm = \mathbf{L}^\pm\dot\sigma^\pm$. These local constitutive rate equations together with the averaging rule and compatibility conditions (8.13) can be solved to yield the constitutive rate-equation for the *averaged* rate-variables, namely

$$\{\dot\sigma\} = \tilde{\mathbf{L}}\{\dot\varepsilon\}, \qquad \{\dot\varepsilon\} = \tilde{\mathbf{M}}\{\dot\sigma\}, \qquad (8.15)$$

where $\tilde{\mathbf{L}}$ and $\tilde{\mathbf{M}}$ are the effective elastic moduli of the laminate, as introduced in Sect. 2.6. This is clear once we note that the governing equations leading to (8.15) are identical to those governing the micro-macro transition of Sect. 2.6 with the local variables replaced by their rates and the transformation strains absent in the local constitutive rate-equations, $\dot\varepsilon^\pm = \mathbf{L}^\pm\dot\sigma^\pm$. It is essential that equations (8.15) are valid irrespective of whether the laminated microstructure is evolving (and the transformation fronts are moving) or not. The effective elastic moduli $\tilde{\mathbf{L}}$ and $\tilde{\mathbf{M}}$ are the instantaneous ones corresponding to the current microstructure which is specified by η, the volume fraction of the product phase.

Using (8.15), the averaged rate-variables $\{\dot{\varepsilon}\}$ and $\{\dot{\sigma}\}$ can be eliminated from (8.12), and the following *macroscopic constitutive rate-equations* for a specified '+' to '−' transformation are obtained

$$\dot{\mathbf{E}} = \tilde{\mathbf{M}}\dot{\boldsymbol{\Sigma}} + \dot{\eta}\,\boldsymbol{\mu}, \qquad \dot{\boldsymbol{\Sigma}} = \tilde{\mathbf{L}}\dot{\mathbf{E}} - \dot{\eta}\,\boldsymbol{\lambda}, \qquad (8.16)$$

where

$$\boldsymbol{\mu} = \tilde{\mathbf{M}}\boldsymbol{\lambda} = \Delta\varepsilon - \tilde{\mathbf{M}}\Delta\sigma. \qquad (8.17)$$

The structure of rate-equations (8.16) resembles that of the well-known constitutive rate-equations of plasticity with $\boldsymbol{\mu}$ and $\boldsymbol{\lambda}$ being the directions of the non-elastic parts of $\dot{\mathbf{E}}$ and $\dot{\boldsymbol{\Sigma}}$, respectively. These directions depend not only on the transformation strains within the layers but also on the elastic properties of the phases. Importantly, the constitutive rate-equations (8.16) are independent of the adopted transformation criterion.

The transformation criterion (8.8) can now be used to determine $\dot{\eta}$ in terms of the rate of the macroscopic strain or stress. During progressive transformation with $\dot{\eta} > 0$, the transformation criterion is continuously satisfied which can be written in the form of the *consistency condition*,

$$\dot{f} - \dot{f}_{\mathrm{c}} = 0 \quad \text{if } \dot{\eta} > 0. \qquad (8.18)$$

In view of uniformity of deformation within each constituent phase, the driving force f (and also $\dot{f}$) is uniform at all interfaces within the laminate, and thus a single value of f (and $\dot{f}$) constitutes, at the same time, the *local* driving force associated with the propagation of each transformation front and the *macroscopic* one associated with the rate $\dot{\eta}$ of the volume fraction of the product phase. Clearly, the consistency condition in the form (8.18) expresses the latter, macroscopic point of view. Note, however, that it is implicitly assumed that the critical driving force f_{c} is also uniform at all interfaces. Note also that, in the case of the local consistency condition, the corresponding time derivative of the thermodynamic driving force must follow the moving phase transformation front, cf. Petryk [102], Stupkiewicz and Petryk [140].

Differentiation of (8.7) yields

$$\dot{f} = \dot{\sigma}^{+} \cdot \Delta\varepsilon - \Delta\sigma \cdot \dot{\varepsilon}^{-}, \qquad (8.19)$$

where the relation $\dot{\phi} = (\partial\phi/\partial\varepsilon) \cdot \dot{\varepsilon} = \sigma \cdot \dot{\varepsilon}$, valid in the case of an isothermal process, has been used. Next, using the compatibility conditions (2.30) and (8.14) and property (2.31), the rates of local variables are replaced by the averaged rate-variables, for example

$$\dot{\sigma}^{+} \cdot \Delta\varepsilon = \dot{\sigma}_{\mathrm{A}}^{+} \cdot \Delta\varepsilon_{\mathrm{A}} = \{\dot{\sigma}_{\mathrm{A}}\} \cdot \Delta\varepsilon_{\mathrm{A}} = \{\dot{\sigma}\} \cdot \Delta\varepsilon, \qquad (8.20)$$

so that

$$\dot{f} = \{\dot{\sigma}\} \cdot \Delta\varepsilon - \Delta\sigma \cdot \{\dot{\varepsilon}\}. \qquad (8.21)$$

Finally, using the constitutive rate-equations (8.15) and definitions (8.17), the rate of the thermodynamic driving force is obtained in the form

$$\dot{f} = \boldsymbol{\lambda} \cdot \{\dot{\varepsilon}\} = \boldsymbol{\mu} \cdot \{\dot{\sigma}\}. \tag{8.22}$$

Using (8.22) and (8.12), the consistency condition (8.18) can now be expressed as

$$\boldsymbol{\lambda} \cdot (\dot{\mathbf{E}} - \dot{\eta}\Delta\varepsilon) = \boldsymbol{\mu} \cdot (\dot{\boldsymbol{\Sigma}} - \dot{\eta}\Delta\sigma) = \dot{f}_c \quad \text{if} \ \dot{\eta} > 0, \tag{8.23}$$

so that $\dot{\eta}$ can be determined in terms of the rates of macroscopic variables. Indeed, introducing quantities $g = \boldsymbol{\lambda} \cdot \Delta\varepsilon$ and $h = \boldsymbol{\mu} \cdot \Delta\sigma$, which in view of (8.17) and (2.31) satisfy

$$g = \boldsymbol{\lambda} \cdot \Delta\varepsilon = \Delta\varepsilon \cdot \tilde{\mathbf{L}}\Delta\varepsilon > 0, \quad h = \boldsymbol{\mu} \cdot \Delta\sigma = -\Delta\sigma \cdot \tilde{\mathbf{M}}\Delta\sigma \leq 0, \tag{8.24}$$

the following macroscopic criterion is obtained from (8.23)

$$\dot{\eta} = \begin{cases} \dfrac{1}{g}(\boldsymbol{\lambda} \cdot \dot{\mathbf{E}} - \dot{f}_c) > 0 \ \text{if} \ \eta < 1 \ \text{and} \ f = f_c \ \text{and} \ \boldsymbol{\lambda} \cdot \dot{\mathbf{E}} > \dot{f}_c, \\ 0 \hspace{3.8cm} \text{otherwise.} \end{cases} \tag{8.25}$$

A dual expression yields

$$\boxed{\dot{\eta} = \dfrac{1}{h}(\boldsymbol{\mu} \cdot \dot{\boldsymbol{\Sigma}} - \dot{f}_c) \quad \text{if} \ \dot{\eta} > 0 \ \text{and} \ \Delta\sigma \neq 0,} \tag{8.26}$$

while

$$\boldsymbol{\mu} \cdot \dot{\boldsymbol{\Sigma}} = \dot{f}_c \quad \text{if} \ \dot{\eta} > 0 \ \text{and} \ \Delta\sigma = 0. \tag{8.27}$$

In a special case of constant f_c, so that $\dot{f}_c = 0$, substitution of (8.25) into (8.16) yields the macroscopic *tangent moduli tensor* $\tilde{\mathbf{L}}^t$ during transformation, and we have

$$\boxed{\dot{\boldsymbol{\Sigma}} = \tilde{\mathbf{L}}^t\dot{\mathbf{E}}, \quad \tilde{\mathbf{L}}^t = \tilde{\mathbf{L}} - \dfrac{1}{g}\boldsymbol{\lambda} \otimes \boldsymbol{\lambda} \quad \text{if} \ \dot{\eta} > 0 \ \text{and} \ f_c = \text{const.}} \tag{8.28}$$

The above macroscopic rate-equations have then exactly the structure of the equations of the classical small-strain elastoplasticity with the normality rule, cf. (3.11). The transformation criterion (8.7) constitutes the "yield" surface and defines the elastic domain, tensors $\boldsymbol{\mu}$ and $\boldsymbol{\lambda}$ are the outward normals to its boundary in the space of macroscopic stress and strain, respectively, and $\dot{\eta}$ plays the role of the plastic multiplier.

Note that a "softening" behaviour during transformation is predicted by (8.26) in a general case of $\Delta\sigma \neq 0$. Indeed, the macroscopic stress increment is then, for $\dot{\eta} > 0$ and $\dot{f}_c = 0$, directed into the elastic domain, $\boldsymbol{\mu} \cdot \dot{\boldsymbol{\Sigma}} < 0$, in view of $h < 0$. If $\Delta\sigma = 0$, then the transformation is associated with "ideal yielding", as predicted by (8.27).

8.2.4 Discussion

Once the critical driving force f_c and its evolution are prescribed, the incremental macroscopic constitutive law of a material undergoing martensitic phase transformation through the evolution of a laminated microstructure is fully described by formulae (8.16) and (8.25)–(8.27). In particular, for a specified macroscopic strain path $\mathbf{E}(t)$, equations (8.16)$_2$ and (8.25) can be integrated to yield both the macroscopic stress response, $\mathbf{\Sigma}(t)$, and the evolution of the microstructure, $\eta(t)$. The local stresses and strains within the layers can then be determined with the help of the micro-macro transition relations provided in Sect. 2.6.

However, as long as the transformation proceeds without elastic unloading or reverse transition, a *path-independent* value of η, corresponding to the current macroscopic strain $\mathbf{E}$, can be found directly from the algebraic condition $f(\mathbf{E}, \eta) - f_c = 0$, where f is expressed in terms of $\mathbf{E}$ and η using the micro-macro transition relations of Sect. 2.6. The numerical simulations presented in the following sections employ the latter approach.

The assumption that the interface normal $\mathbf{n}$ is constant during transformation is an essential element of the present modelling. The problem of selecting the initial orientation of martensitic plates, which does not change during transformation in view of the above assumption, has not been addressed yet. As a first approximation, the interface normal following from the crystallographic theory is adopted in the examples below. Another approach, which is, in fact, a direct consequence of the phase transformation criterion (8.8), is developed in Chap. 9.

By ruling out the microstructural changes within the layers, the analysis of Sect. 8.2.3 has been restricted to the case of evolving rank-one laminates. The present micromechanical approach is, however, more general and, under the assumption of separation of scales, can be applied to higher-rank laminated microstructures. The micro-macro transition of Sect. 2.6 is then carried out at the lowest level of a nested laminated microstructure. The resulting effective properties are used at the next level, and the procedure is repeated. Finally, the relevant transformation criteria, expressed in terms of the respective thermodynamic driving forces, are formulated using the local variables (stresses and strains) at each microstructural level.

8.3 Evolving Rank-One Laminate

8.3.1 Computational Scheme

In this section, the micromechanical model and the corresponding computational scheme are specified for the case of stress-induced transformation proceeding through nucleation and growth of parallel martensitic plates, thus an evolving *rank-one laminate* is considered. Furthermore, the *path-independent* formulation is only addressed, so that the transformation is assumed to proceed without elastic unloading or reverse transformation.

As a first approximation, we assume that the microstructural parameters of martensitic plates follow from the crystallographic theory of martensite, cf. Sect. 7.2, here in the geometrically linear setting which is consistent with the present micromechanical framework. This is an approximation since the crystallographic theory assumes stress-free conditions and thus zero elastic strains. This is clearly not the case in stress-induced transformations. However, the assumption can be regarded acceptable because the elastic strains are small compared to the transformation strains. An extension of the crystallographic theory, that accounts for the elastic strains, is presented in Chap. 9.

For a given alloy and for specified parent and product phases,[3] the crystallographic theory provides the microstructural parameters of N distinct martensitic plates. The martensitic plates are indexed by $\alpha = 1, \ldots, N$, and the essential microstructural parameters for each plate are the habit plane normal $\mathbf{m}_\alpha$ and the effective transformation strain $\hat{\varepsilon}^t_\alpha$, cf. Sect. 7.2. The effective transformation strain of an internally twinned martensite is given by (7.12) and that of an internally faulted martensite by (7.14).

Consider now the isothermal pseudoelastic response of a single crystal of a shape memory alloy under a given loading program. The temperature is assumed sufficiently high for the austenite to be stable at zero macroscopic stress. The loading program is prescribed by providing six equations for the components of macroscopic stress $\mathbf{\Sigma}$ or strain $\mathbf{E}$ in terms of a single control parameter p which is assumed to increase monotonically. The thermodynamic driving force $f^{A \to M}_\alpha = f_\alpha$ can then be computed from (8.7) and (2.37)–(2.38) assuming that parallel martensitic plates of the same family, indexed by α, form during the austenite-to-martensite (A→M) transformation,

$$f_\alpha = \hat{f}_\alpha(\eta, p) = \mathbf{\Sigma} \cdot \hat{\varepsilon}^t_\alpha - \frac{1}{2} \mathbf{\Sigma} \cdot \mathbf{B}^T_a (\mathbf{M}_a - \mathbf{M}_\alpha) \mathbf{B}_\alpha \mathbf{\Sigma} - \Delta^{am} \phi_0, \qquad (8.29)$$

where the austenite is identified with the '+' phase and the α-th martensitic plate with the '−' phase. In (8.29), $\mathbf{B}_a$ and $\mathbf{B}_\alpha$ are the stress concentration tensors of austenite and martensite, respectively, and $\mathbf{M}_\alpha$ is the elastic compliance tensor of the martensitic plate (effective one in the case of twinned martensite). In the case of reverse transformation, the physical meaning of '+' and '−' phases is opposite and thus the respective thermodynamic driving force is $f^{M \to A}_\alpha = -f^{A \to M}_\alpha = -f_\alpha$. Note that the expression for f_α, specified by (8.29), corresponds to a *compatible* transformation strain $\hat{\varepsilon}^t_\alpha$. In a general case, additional terms related to incompatibility of the transformation strain appear in the expression for f_α. The explicit expression for f_α in that general

---

[3] In some alloys, several martensitic phases exist and multiple austenite–martensite transformations are possible. Moreover, successive transformations between different martensitic phases are also observed, e.g. Otsuka et al. [92], Šittner et al. [155].

case is not provided here, but it can be easily derived using the micro-macro transition relations for simple laminates, Sect. 2.6.

Initially, the crystal is in austenitic state ($\eta = 0$) and, as long as $f_\alpha < f_c$ for all α, the austenite is loaded elastically. The threshold driving force f_c is assumed identical for all martensitic plates and constant during transformation. As the control parameter p changes, the transformation condition is checked for all possible martensitic plates and the *preferred plate* is chosen for which the transformation criterion $f_\alpha = f_c$ is satisfied first. Further loading results in progressive transformation with increasing volume fraction η of the preferred plates. The relation between η and p is then implicitly defined by the transformation condition $f_\alpha = f_c$, as long as $\dot{\eta} > 0$. The response is elastic after the transformation is completed ($\eta = 1$) which corresponds to elastic loading of the martensite. Reverse transformation is treated analogously and governed by $f_\alpha = -f_c$ with $\dot{\eta} < 0$. Elastic unloading is also possible when the material is only partially transformed ($0 < \eta < 1$), and the elastic domain is bounded by $-f_c \leq f_\alpha \leq f_c$.

---

1. For each martensitic plate α, find the control parameter p_α^0 for which the respective transformation condition is satisfied in a purely austenitic state: solve $\hat{f}_\alpha(0, p_\alpha^0) = f_c$ with $\hat{f}_\alpha(\eta, p)$ given by (8.29).
2. Choose the preferred martensitic plate for which the transformation is initiated first, i.e. that corresponding to the smallest value of p_α^0 (the control parameter p is assumed to increase monotonically).
3. Compute the relationship between the volume fraction η of the preferred martensitic plates ($0 \leq \eta \leq 1$) and the control parameter p from the transformation condition $\hat{f}_\alpha(\eta, p) = f_c$ with $\hat{f}_\alpha(\eta, p)$ given by (8.29).
4. Compute the elastic response of pure austenite (before the transformation is initiated, $\eta = 0$) and pure martensite (after transformation is completed, $\eta = 1$).
5. Using the formulae given in Sect. 2.6 and the calculated relationship between η and p, determine the overall and local stresses and strains in the transformation process.

Box 8.1

The basic steps involved in the simulation of the forward austenite-to-martensite transformation are summarized in Box 8.1. Upon unloading, the reverse transformation is treated accordingly, however, only the initially preferred martensitic plates, i.e. those selected during the forward transformation, are considered.

As a consequence of the adopted micromechanical approach, the material parameters involved in the model have a clear physical meaning. The transformation strains of martensite variants are known from the crystallography of transformation: they are computed using the lattice parameters of parent and product phases. The effective transformation strains of martensitic plates follow then from the crystallographic theory, cf. Sect. 7.2. The elastic anisotropy of the phases is fully accounted for in the model, thus also the elastic moduli tensors of single crystals in austenitic and martensitic states are required. These parameters, though rather fundamental, are not easily found in the literature, and for the majority of alloys and transformations they are, in fact, not available. Finally, the model involves the chemical energy $\Delta^{am}\phi_0$ and the threshold driving force f_c. The former one is a linear function of temperature and can be identified, for instance, from the temperature dependence of the transformation stress, i.e. the stress at which the transformation initiates. The last parameter, f_c, is directly related to the hysteresis in the isothermal closed forward–reverse transformation cycles, as discussed in Sect. 8.2.2, and can be identified from the width of hysteresis loops.

Remark 8.2. The above set of material data fully characterizes the material. However, a complete set of material parameters is rarely available, the elastic constants of single-crystalline martensite being the critical parameters, most difficult to find. In fact, the elastic constants of both the austenite and the martensite are, at the moment, available only for two transformations (the cubic-to-monoclinic transition in CuZnAl and the cubic-to-orthorhombic transition in CuAlNi), and the numerical examples provided in this chapter, and in Chap. 9, concern mostly these two transformations.

Remark 8.3. The transformation condition $f_\alpha = \pm f_c$ is the basic governing equation of the model. The term $\boldsymbol{\Sigma} \cdot \hat{\varepsilon}_\alpha^t$ in the expression for f_α in (8.29) is recognized as the so-called *Schmid factor* (by the analogy to single-crystal plasticity). The second term in (8.29), quadratic in the macroscopic stress $\boldsymbol{\Sigma}$, accounts for the residual stresses associated with the difference in elastic properties of the two phases; note the term $\mathbf{M}_a - \mathbf{M}_\alpha$. The present macroscopic transformation criterion is thus different from the widely used extension of the Schmid law, e.g. Shield [126], Šittner and Novák [154], Thamburaja and Anand [149]. Although, as illustrated by the examples below, the correction with respect to the Schmid law is usually not essential, the present model provides additional information such as redistribution of local stresses within the layers. Furthermore, the model can easily be extended to the case of incompatible transformation strains, evolving higher-rank laminates, etc., as discussed in Sect. 8.4 and in Chap. 9.

Remark 8.4. In the elastically homogeneous case ($\mathbf{M}_a = \mathbf{M}_\alpha$) and for compatible transformation strain $\hat{\varepsilon}_\alpha^t$, the local stresses are continuous within the laminate (and thus equal to the macroscopic one), and the transformation criterion $f_\alpha = f_c$, which follows from the present model, reduces to the Schmid law.

8.3.2 Untwinned Martensite in CuAlNi Alloy

As the first application of the model outlined above, we consider the pseudo-elastic response associated with the stress-induced cubic-to-monoclinic transition in a CuAlNi single crystal. In this transition the cubic austenite of DO_3 structure (β_1 phase) transforms to the monoclinic martensite of 6M structure (β_1' phase), cf. Otsuka et al. [91]. The martensite appears in the form of untwinned plates and the compatibility at the austenite–martensite interface is provided by random stacking faults on the basal (101) planes (internally faulted martensite). More details concerning the crystallography and the mechanism of the transformation are provided in Chap. 9.

The cubic-to-monoclinic transformation gives rise to 12 variants of martensite. In Cu-based alloys, these variants have the "cubic axes" structure with a unique twofold axis along the edge of the original cubic unit cell, cf. Pitteri and Zanzotto [107]. For each of these variants, the habit plane equation (7.14) has two solutions, there are thus $N = 12 \times 2 = 24$ distinct, but crystallographically equivalent, martensitic plates. The numerical values of the microstructural parameters (habit plane normal $\mathbf{m}$, shape strain vector $\mathbf{b}$, and shear magnitude due to stacking faults k_{sf}) are given in Appendix B. Note that the transformation strain of a compatible martensitic plate is sufficiently characterized by vectors $\mathbf{m}$ and $\mathbf{b}$, cf. (7.13) and (7.14).

Elastic constants of the cubic β_1 austenite are adopted from the literature, cf. Suezawa and Sumino [143]. Since the elastic constants of the monoclinic β_1' martensite are not available in the literature, in the computations below the elastic properties of the β_1' phase are estimated by scaling the elastic constants of a similar martensitic phase in a CuZnAl alloy. This is discussed in more detail in Appendix B, where also the values of the elastic constants are provided.

Two more parameters are required to completely characterize the material: the chemical energy $\Delta^{am}\phi_0$ and the critical thermodynamic driving force f_c. Roughly speaking, the chemical energy $\Delta^{am}\phi_0$ is related to the stress at which transformation proceeds (at a given temperature) and the critical driving force f_c is related to the width of hysteresis loop. In general, these parameters depend on the alloy composition and on heat treatment. Their values are identified below for specific experimental data.

Experiments of Horikawa et al. [48]

Horikawa et al. [48] reported complete stress–strain diagrams of eight differently oriented single-crystal specimens of CuAlNi. Stereographic projection of the corresponding tension axis orientations, relative to the cubic basis of the austenite, are provided in Fig. 8.4(a). Figure 8.5 presents the orientation dependence of stress–strain response predicted by the present model, compared to the experimental data (dashed lines). The values of the chemical energy $\Delta^{am}\phi_0 = 12\,\text{MJ/m}^3$ and of the critical driving force $f_c = 0.4\,\text{MJ/m}^3$ have been adjusted to correctly represent the initial transformation stress and the hysteresis width of specimen '1'. The remaining predictions are obtained using the same set of parameters.

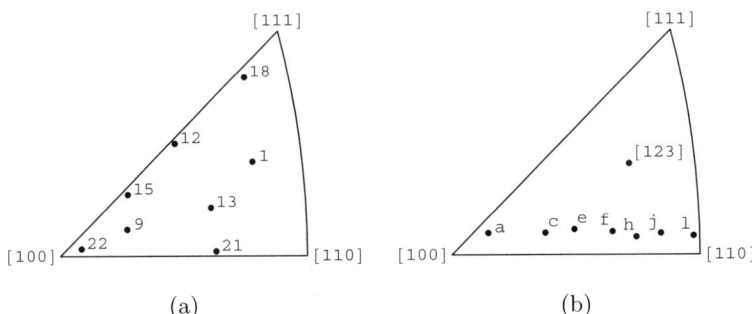

Fig. 8.4. Load axis orientations of CuAlNi single crystal specimens used by Hori-kawa et al. [48] (a) and Novák et al. [86] (b). The orientations are indicated in the unit triangle, relative to the cubic basis of austenite.

Both the experimental and the theoretical stress–strain curves exhibit sub-stantial orientation-dependence in terms of the inelastic transformation strain, transformation stress, and hysteresis width. With respect to all these aspects, the agreement of the model predictions with the experimental data is satis-factory. As an exception, the experimentally observed transformation stress of specimen '21' is significantly higher than that predicted by the model.[4] *The general rule is that the smaller the transformation strain the higher the transformation stress and the wider the hysteresis loop.* The latter observation is consistent with the result which stems from the adopted transformation cri-terion, namely that the area of the hysteresis loop on a stress–strain diagram corresponding to a complete forward–reverse transformation is constant and equal to $2f_c$, cf. (8.9).

Elastic anisotropy of the austenite,[5] prior to transformation, is also pro-perly described. Only in the case of specimen '22' the measured elastic stiffness is about twice smaller than predicted. The discrepancy between the measured

---

[4] The experimental transformation stress of specimen '21' seems to be inconsistent with the remaining experimental data. In particular, the observed transforma-tion strain of specimens '9', '13', and '15' is also between 8 and 9 per cent, and the corresponding transformation stresses are about 150–160 MPa, while that of specimen '21' is nearly 250 MPa. A possible explanation of this discrepancy is that the reported stress–strain response of specimen '21' corresponds to a higher temperature.

[5] The cubic β_1 phase in CuAlNi exhibits very high elastic anisotropy. This is reflec-ted by the value of the anisotropy index $A = 2c_{44}/(c_{11} - c_{12})$ as high as $A = 12$ for the elastic constants of the β_1 phase, cf. Table B.3.

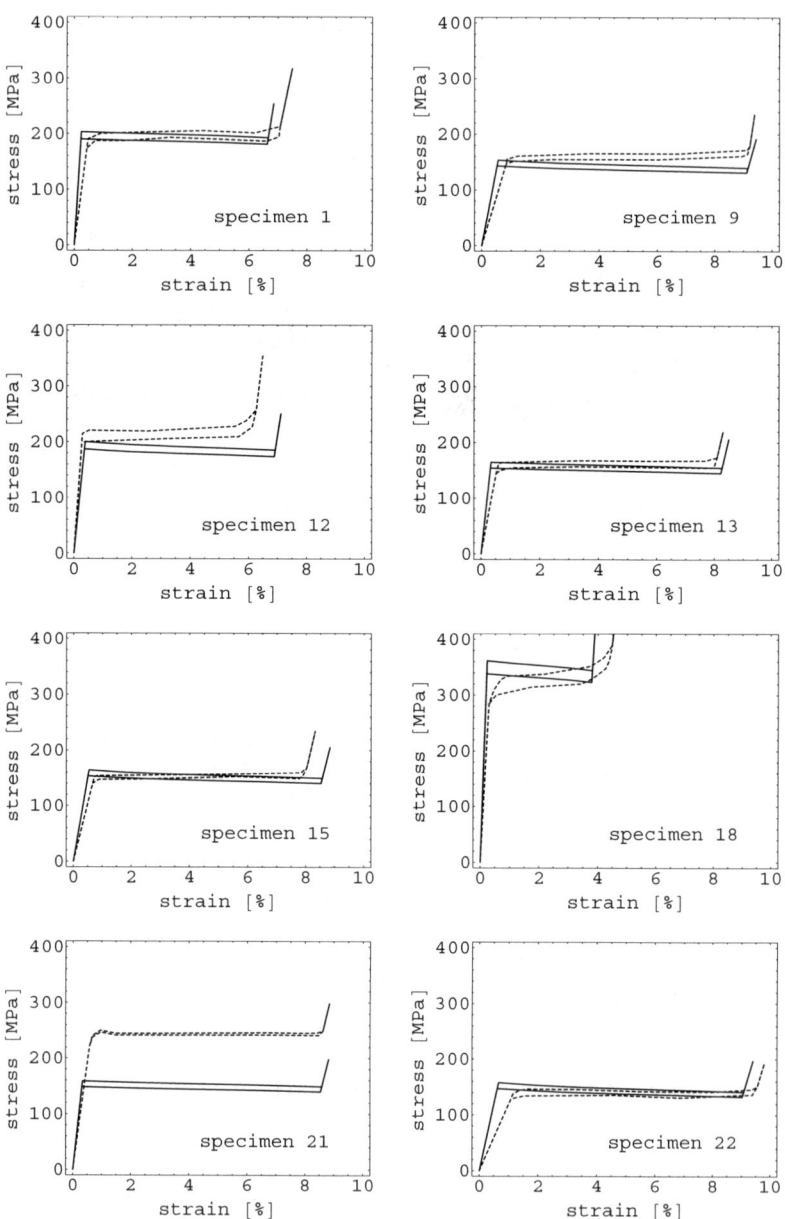

Fig. 8.5. Orientation dependence of stress–strain response in uniaxial tension of CuAlNi single crystals undergoing the $\beta_1 \rightarrow \beta_1'$ transformation: model predictions (solid lines) and experimental results of Horikawa et al. [48] (dashed lines).

and predicted elastic stiffness of specimen '22' has already been noticed by Horikawa et al. [48].

As shown in Sect. 8.2.3, the present model predicts a softening behaviour during progressive transformation provided that the local stresses within the layers are discontinuous at the transformation front, $\Delta\boldsymbol{\sigma} \neq \mathbf{0}$. In the analyzed case, although the transformation strain is compatible, the local stresses are not homogeneous because of the elastic mismatch of the phases ($\Delta\mathbf{M} \neq \mathbf{0}$), cf. $(2.33)_2$. Accordingly, the softening behaviour is clearly visible in Fig. 8.5, as predicted by the theory, cf. (8.26).

The effect of the elastic mismatch is further illustrated in Fig. 8.6 where the uniaxial stress–strain diagrams corresponding to different temperatures are provided for two representative specimen orientations. The effect of temperature on the isothermal pseudoelastic response is accounted for by expressing the chemical energy $\Delta^{\mathrm{am}}\phi_0$ as a linear function of temperature, cf. (8.5); other effects, e.g. thermal expansion, are neglected. In Fig. 8.6, T_0 corresponds to the diagrams presented in Fig. 8.5, $\Delta T = 38.5\,\mathrm{K}$, and $\Delta^{\mathrm{am}}s_0 = 0.156\,\mathrm{MJ/m^3K}$, cf. Sect. 8.3.4.

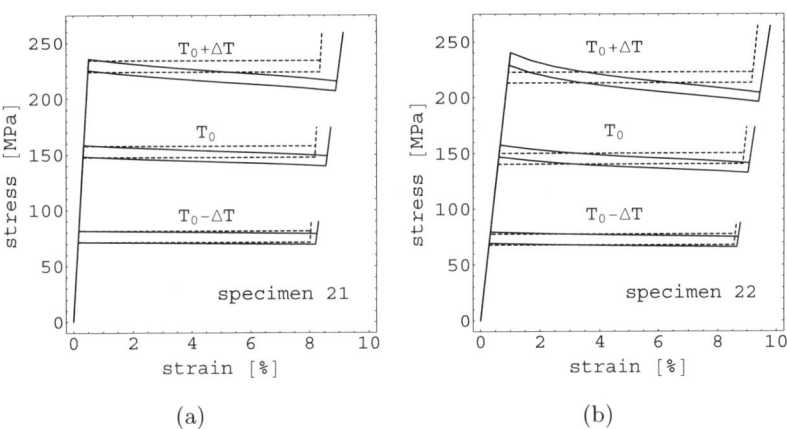

Fig. 8.6. Stress–strain diagrams in uniaxial tension of CuAlNi single crystals at different temperatures. Dashed lines denote the prediction of the Schmid law (elastically homogeneous case, $\mathbf{M}_\alpha = \mathbf{M}_a$).

The elastic mismatch term in the expression for the driving force f_α is quadratic in $\boldsymbol{\Sigma}$, cf. (8.29). Thus the corresponding effects (softening and deviation from the prediction of the Schmid law) are more pronounced when the transformation stress is higher, i.e. at higher temperatures. This is seen in Fig. 8.6. The transformation stress predicted by the Schmid law, cf. dashed lines in Fig. 8.6, is given by the following formula

$$\Sigma^{\mathrm{Schmid}} = \frac{\Delta^{\mathrm{am}}\phi_0 \pm f_c}{(\mathbf{t} \otimes \mathbf{t}) \cdot \hat{\boldsymbol{\varepsilon}}_\alpha^{\mathrm{t}}}, \tag{8.30}$$

which follows from (8.29) for $\mathbf{M}_\alpha = \mathbf{M}_a$. Here, Σ^{Schmid} is the uniaxial transformation stress, $\mathbf{t}$ is the orientation of the load axis, and the plus (minus) sign in (8.30) corresponds to the forward (reverse) transformation.

Experiments of Novák and Šittner

Single crystals of CuAlNi under compressive loading have been investigated by Novák et al. [86]. It has been observed that, depending on the specimen orientation, either the monoclinic β_1' phase or the orthorhombic γ_1' phase was induced. The cubic-to-monoclinic $\beta_1 \rightarrow \beta_1'$ transformation dominated for load axis orientations closer to the [011] pole (specimens 'h', 'j', and 'l') while the cubic-to-orthorhombic $\beta_1 \rightarrow \gamma_1'$ transformation dominated for load axis orientations closer to the [001] pole (specimens 'a', 'c', 'e', and 'f'). The load axis orientations are shown in Fig. 8.4(b).

In Fig. 8.7(a,b), the stress–strain diagrams predicted by the present model for specimens 'h' and 'l' are compared to the experimental results of Novák et al. [86]. In the present computations, the transformation strains and the elastic constants are those used above to simulate the results of Horikawa et al. [48], while the chemical energy $\Delta^{\mathrm{am}}\phi_0 = 19.5\,\mathrm{MJ/m^3}$ and the critical driving force $f_c = 1.2\,\mathrm{MJ/m^3}$ have been roughly identified to match the experimental transformation stress and hysteresis width of specimen 'l'.

In order to illustrate the tension-compression asymmetry, the pseudoelastic response of a CuAlNi single crystal in tension and compression is also provided in Fig. 8.7(c). The experimental results are taken from Šittner and Novák [154] and refer to the same alloy as that used by Novák et al. [86]; the load axis orientation is [123], cf. Fig. 8.4(b).

It is seen from Fig. 8.7 that the agreement with experimental data is not as good as in the case of experiments of Horikawa et al. [48], cf. Fig. 8.5. The observed transformation strains and the apparent elastic moduli of austenite are significantly smaller than those predicted by the model, so is the hysteresis width in the case of the [123] specimen. However, the variation of the transformation stress with the load axis orientation and with the sense of loading is properly predicted.

According to Šittner [152], the experimentally observed elastic moduli are too small (compared to the moduli consistent with the elastic properties measured using the ultrasonic technique, e.g. Suezawa and Sumino [143], Landa et al. [76]) due to the softness of the testing machine. However, the tendency that the apparent Young's modulus is lowest for [001] orientations and increases towards [011] and [111] orientations is reflected in the experimental results. The reason for the discrepancy between the observed and the predicted transformation strains is not known. A possible explanation could be that the phase transformation was not complete within the specimen.

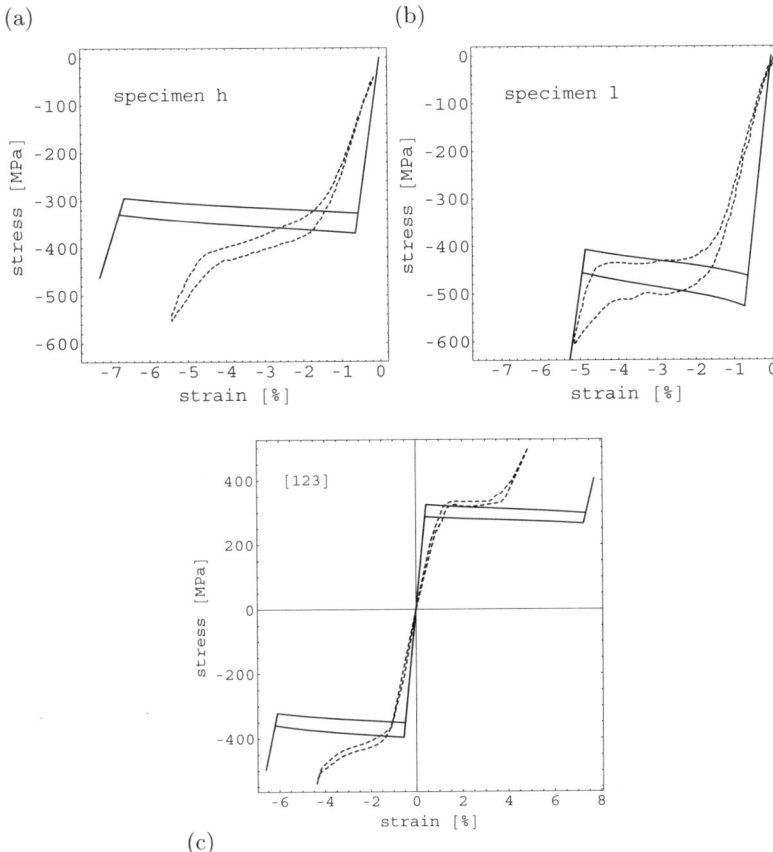

Fig. 8.7. Stress–strain diagrams in uniaxial tension and compression of CuAlNi single crystals undergoing the $\beta_1 \rightarrow \beta_1'$ transformation: model predictions (solid lines) and experimental results of Novák et al. [86] and Šittner and Novák [154].

8.3.3 Twinned Martensite in CuAlNi Alloy (No Detwinning)

In the model outlined in Sect. 8.3.1, it is assumed that the martensitic plates do not undergo any *internal* microstructural changes during transformation. Thus direct application of the model for internally twinned martensites is only possible if martensite variant rearrangement (detwinning) is neglected. This assumption may be non-physical because the twin boundaries are highly mobile and detwinning is expected to accompany the austenite-to-martensite transformation. In this section, detwinning is neglected, and the corresponding predictions are provided for uniaxial compression of CuAlNi single crystals. The extension of the model to account for detwinning is presented in Sect. 8.4.

In the case of internally twinned martensites, the microstructure corresponding to the formation and growth of parallel martensitic plates is a rank-two laminate, cf. Fig. 8.2(a). Assuming that the internal structure of each martensitic plate is fixed, i.e. the twin fraction λ is constant, the transformation is associated with microstructural changes only at the level of austenite–martensite laminate. The effective properties of martensitic plates, which are then constant during the transformation, are obtained by applying the micro–macro transition scheme at the level of twinned martensite. The model of Sect. 8.3.1 and the expression (8.29) for the thermodynamic driving force at the austenite–martensite interface are thus directly applicable with $\mathbf{M}_\alpha$ and $\hat{\varepsilon}_\alpha^t$ being, respectively, the effective elastic moduli tensor and the effective transformation strain of an internally twinned martensitic plate.

As an example, we consider the cubic-to-orthorhombic $\beta_1 \rightarrow \gamma_1'$ transformation in CuAlNi. The stress–strain diagrams of differently oriented single crystals in uniaxial compression are compared to the experimental results of Novák et al. [86] in Fig. 8.8. The load axis orientations relative to the cubic lattice of the austenite are shown in Fig. 8.4(b). Only specimens 'a' to 'f' are analyzed, for which the $\beta_1 \rightarrow \gamma_1'$ transformation was observed[6] by Novák et al. [86]. Model parameters (elastic constants and transformation strains) are provided in Appendix B, the remaining parameters, $\Delta^{am}\phi_0 = 9.3\,\mathrm{MJ/m^3}$ and $f_c = 2\,\mathrm{MJ/m^3}$, have been adjusted using the stress–strain curve of specimen 'a', cf. Fig. 8.8(a).

As in the case of the $\beta_1 \rightarrow \beta_1'$ transformation, cf. specimens 'h' and 'l' in Fig. 8.7, the agreement with the experimental data is not satisfactory in some respects. Contrary to the case of the $\beta_1 \rightarrow \beta_1'$ transformation, in the present case, the predicted transformation strains are significantly smaller that those observed in the experiments. The additional strain can be attributed to detwinning which is by assumption ruled out in the present simulations. This is further discussed in Sect. 8.4.

The discrepancy of the apparent elastic moduli of the austenite has already been discussed in Sect. 8.3.2. The predicted orientation dependence of the transformation stress agrees reasonably well with the experimental dependence. The hysteresis width predicted for specimen 'c' is in agreement with experiment (note that f_c was adjusted for specimen 'a'), but in the case of specimens 'f' and 'h' an additional growth of hysteresis width is observed, also the character of the stress–strain response is different. A possible reason is that a more complex transformation pattern develops in specimens 'f' and 'h', with multistage or successive $\beta_1 \rightarrow \beta_1' \rightarrow \gamma_1'$ transformations, cf. Kato et al. [58] and Šittner et al. [155].

It is seen from Fig. 8.8 that, for the internally twinned martensites, the model predictions without detwinning have essentially the same character as in the case of untwinned martensites: the transformation proceeds at a nearly

---

[6] It is commonly agreed that the width of the hysteresis loop in the stress–strain diagram is a good indicator of the type of transformation in CuAlNi alloys: the hysteresis associated with the $\beta_1 \rightarrow \beta_1'$ transformation is much smaller than that of the $\beta_1 \rightarrow \gamma_1'$ transformation, compare Figs. 8.7 and 8.8.

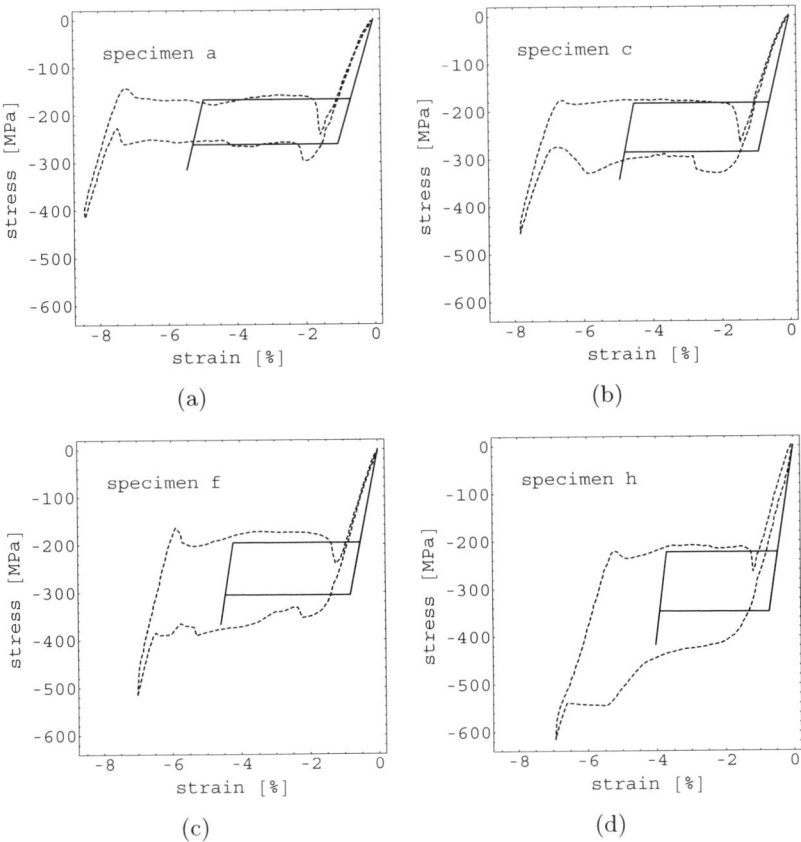

Fig. 8.8. Stress–strain diagrams in uniaxial compression of CuAlNi single crystals undergoing the $\beta_1 \rightarrow \gamma_1'$ transformation: model predictions (no detwinning, solid lines) and experimental results of Novák et al. [86] (dashed lines).

constant stress with some softening due to elastic mismatch. The behaviour is very different when detwinning is included in the model, as discussed in Sect. 8.4. More results concerning the $\beta_1 \rightarrow \gamma_1'$ transformation in CuAlNi, proceeding without detwinning, can be found in Stupkiewicz and Petryk [140].

8.3.4 Macroscopically Adiabatic Case

A simple extension of the present model to the macroscopically adiabatic case is presented below, based on the assumption that the temperature is homogeneous at the micro-scale and equal to the macroscopic one which changes during transformation due to the latent heat of transformation. Thermal expansion and variation of elastic properties with temperature are neglected in the present simple description of thermal effects.

The main effect that is expected in the adiabatic case is the temperature-driven *increase* of the stress during transformation. This is because the martensitic transformation is exothermic. Thus, in the adiabatic conditions, the temperature increases as the transformation proceeds, which results in the increase of the chemical energy $\Delta^{am}\phi_0$ and, consequently, in the increase of the transformation stress. The adiabatic case provides thus an upper bound of the related thermal effects. Clearly, the reverse martensite-to-austenite transformation is endothermic, so that, during the reverse transformation, the stress is expected to decrease.

In macroscopically adiabatic conditions, the heat balance of the r.v.e. undergoing austenite-to-martensite transformation can be written as

$$\rho c\,\mathrm{d}T = \Delta s^*\,T\,\mathrm{d}\eta + f_c\,\mathrm{d}\eta, \qquad \mathrm{d}\eta > 0, \tag{8.31}$$

where the entropy change of transformation $\Delta s^* = s_a^0 - s_m^0 > 0$ is a material parameter, cf. Sect. 8.2.1, $\Delta s^*\,T$ is the latent heat of transformation per unit transformed volume, $f_c\,\mathrm{d}\eta$ is the heat due to intrinsic dissipation, and ρc is the specific heat per unit volume (assumed identical in the austenite and martensite). Equation (8.31) can be obtained from equation (21) in Raniecki and Lexcellent [112] by neglecting the heat due to the piezocaloric effect and the term related to temperature-dependence of the configurational energy associated with micro-stresses.

Evolution of temperature during the forward transformation ($\dot{\eta} > 0$) can thus be determined from the following differential equation,

$$\rho c\,\frac{\mathrm{d}T}{\mathrm{d}\eta} = \Delta s^*\,T + f_c, \qquad T(0) = T_a, \tag{8.32}$$

which upon integration yields the temperature as a function of the volume fraction η of martensite,

$$T^{(f)} - T_a = \left(T_a + \frac{f_c}{\Delta s^*}\right)(e^{\eta\,\Delta s^*/\rho c} - 1), \tag{8.33}$$

where T_a is the initial temperature of austenite (at $\eta = 0$). The expression for the case of reverse transformation

$$T^{(r)} - T_m = \left(T_m - \frac{f_c}{\Delta s^*}\right)(e^{(\eta-1)\Delta s^*/\rho c} - 1), \tag{8.34}$$

is obtained from the heat balance equation, $\rho c\,\mathrm{d}T = \Delta s^*\,T\,\mathrm{d}\eta - f_c\,\mathrm{d}\eta$, corresponding to $\mathrm{d}\eta < 0$. Here T_m is the temperature of martensite at the onset of reverse transformation (at $\eta = 1$). Finally, the variation of the chemical energy in a macroscopically adiabatic transformation process is given by (8.5).

Two limiting cases can now be considered. In a fully adiabatic process, the temperature after a complete forward–reverse transformation cycle slightly increases due to the dissipation associated with non-zero critical driving force

$f_c > 0$, cf. Fig. 8.9(a). In the second case, after the adiabatic forward transformation process, the material is cooled down to the initial temperature, and
then the reverse transformation is induced upon unloading again in adiabatic
conditions, so that $T_m = T_a$. The corresponding evolution of temperature is
sketched in Fig. 8.9(b).

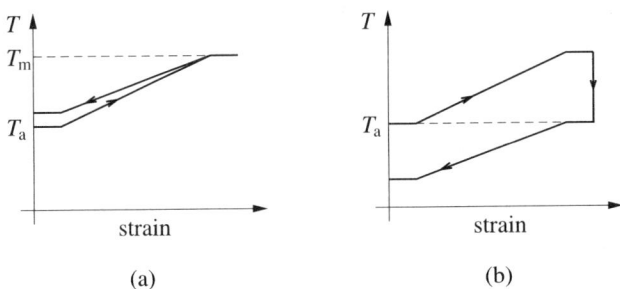

(a)   (b)

Fig. 8.9. Temperature evolution during the stress-induced transformation in adiabatic conditions: (a) fully adiabatic case, (b) adiabatic loading–cooling–adiabatic
unloading.

Consider thus uniaxial tension of a CuAlNi single crystal undergoing the
$\beta_1 \rightarrow \beta_1'$ transformation. The model involves two additional material parameters which are identified using the experimental data of Rodriguez and
Brown [115]. The specific heat is $\rho c = 3.1\,\mathrm{MJ/m^3K}$, and the entropy change
of the transformation[7] is $\Delta s^* = 0.156\,\mathrm{MJ/m^3K}$. The remaining material parameters are those used in Sect. 8.3.2 to simulate the experiments of Horikawa
et al. [48].

The stress–strain diagrams corresponding to the two cases discussed above
are shown in Fig. 8.10 for the initial temperature $T_a = 308\,\mathrm{K}$ and for three load
axis orientations (specimens '1', '9', and '18'). In the present case, the temperature is a nearly linear function of the volume fraction η. This is because the
term $\Delta s^*/(\rho c) = 0.0503$ in (8.33) is rather small. The increase of temperature
associated with the complete forward transformation is predicted to be equal
to 16.03 K which is in a perfect agreement with the temperature increase measured by Rodriguez and Brown [115] in a nearly adiabatic, high strain-rate
stress-induced transformation. At the same time, the predicted temperature

---

[7] The entropy change of the transformation is determined from the formula $\Delta s^* = \Delta S_{\beta_1'-\beta_1}/V_m$, where $\Delta S_{\beta_1'-\beta_1} = 1.20\,\mathrm{J/mol\,K}$ is the entropy change per mole and $V_m = 7.7 10^{-6}\,\mathrm{m^3/mol}$ is the molar volume, both parameters have been provided by Rodriguez and Brown [115].

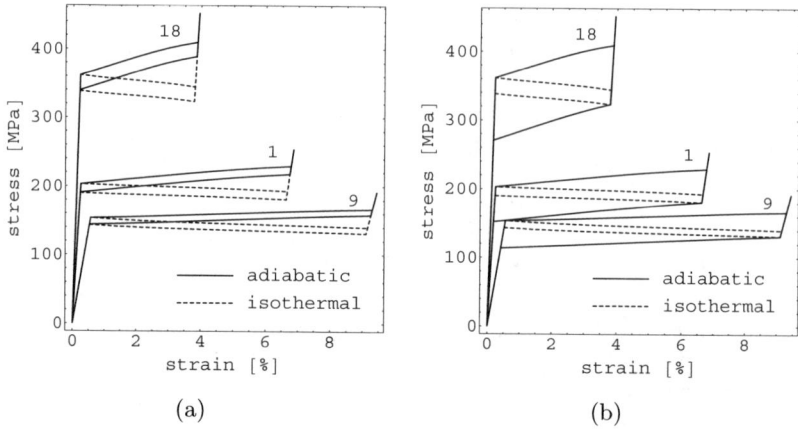

Fig. 8.10. Adiabatic stress–strain diagrams for three load axis orientations in uniaxial tension of CuAlNi single crystal: (a) fully adiabatic case, (b) adiabatic loading–cooling–adiabatic unloading, cf. Fig. 8.9.

increase[8] after the adiabatic forward–reverse transformation cycle is equal to $0.252\,\mathrm{K}$.

The present analysis illustrates two effects expected to appear in non-isothermal processes: apparent hardening due to latent heat and the increase of hysteresis width in the stress–strain diagram due to heat conduction. The latter effect is modelled here by assuming a cooling segment between adiabatic loading and unloading. Both effects have been observed experimentally by Lexcellent et al. [79] who also proposed a simple model accounting for the thermal effects in non-adiabatic conditions. The present model could easily be extended to non-adiabatic conditions by applying the approach of Lexcellent et al. [79], i.e. by introducing the convective heat exchange with the environment, and by assuming that heat conduction within the specimen is immediate.

8.4 Transformation and Detwinning: Evolving Rank-Two Laminate

8.4.1 Mobile Twin Interfaces

In the case of internally twinned martensites, a martensitic plate is a fine mixture of two twin-related martensite variants, cf. Fig. 8.11. Contrary to the case analyzed in Sect. 8.3.3, we now assume that the internal microstructure

---

[8] Neglecting the temperature variation of the latent heat, an estimate of this increase of temperature is given by $2f_c/(\rho c)$, which gives $0.258\,\mathrm{K}$ in the present case.

may change during transformation due to migration of twin boundaries. This is associated with martensite variant rearrangement (detwinning) and with the variation of effective properties (transformation strain and elastic moduli tensor) of the martensitic plates during transformation.

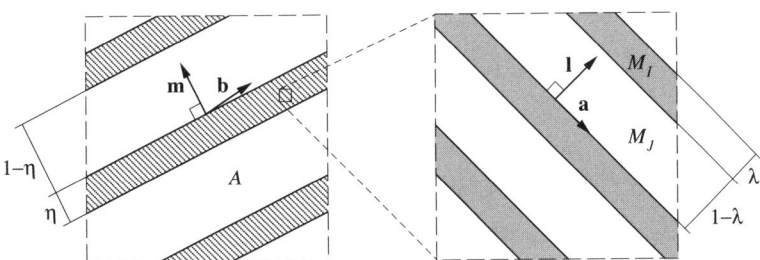

Fig. 8.11. Microstructure (rank-two laminate) associated with the formation of parallel, internally twinned martensitic plates (originally published in [141], copyright EDP Sciences).

Within the present modelling approach, a mobile twinning plane is treated as a phase transformation front with zero chemical energy associated with the transformation from one martensite variant to another, cf. (8.4). Furthermore, we assume for simplicity that the critical driving force for propagation of a twinning plane is also equal to zero. This implies that there is no dissipation directly associated with detwinning. This assumption can easily be relaxed to allow some, typically small, dissipation related to propagation of twin interfaces. However, this has not been found to change the results significantly.

Assuming that the transformation proceeds by formation and growth of parallel martensitic plates, the microstructure at hand is an evolving rank-two laminate. This microstructure is fully characterized once the following parameters are determined: martensite variant pair (I, J), twinning plane and habit plane normal vectors, $\mathbf{l}$ and $\mathbf{m}$, respectively, the twin fraction λ, and the volume fraction of martensite η, cf. Fig. 8.11.

Consider now the forward transformation from austenite to martensite, and assume first that the variant pair (I, J) and vectors $\mathbf{l}$ and $\mathbf{m}$ are known. For given external loading conditions, the *path-independent* values of the remaining two microstructural parameters, λ and η, can be found from two equations: the transformation criterion at the twinning plane, $f^{JI} = 0$, and the transformation criterion at the austenite-twinned martensite plate interface, $f^{am} = f_c^{am}$. Here, f^{JI} and f^{am} denote the respective thermodynamic driving forces, f_c^{am} is the critical driving force for the austenite-to-martensite transformation, and zero critical driving force for the propagation of twinning planes is assumed, $f_c^{JI} = 0$. The local stresses and strains necessary to compute the driving forces f^{JI} and f^{am} follow from the micro-macro transition formulae for the rank-two laminate at hand. These formulae are obtained by applying sequentially the micro-macro transition relationships of simple laminates, Sect. 2.6.

The variant pair (I, J) and vectors $\mathbf{l}$ and $\mathbf{m}$ are selected by applying the above procedure at $\eta = 0$, i.e. at the onset of transformation, for all microstructures (i.e. all possible martensite plates) predicted by the crystallographic theory. The preferred martensite plate is then chosen for which the transformation is initiated first for the prescribed loading program, just like in the case of untwinned martensites, cf. the computational scheme presented in Sect. 8.3.1.

8.4.2 Uniaxial Tension of CuAlNi Single Crystal

As an example, consider isothermal uniaxial tension of a CuAlNi single crystal undergoing the cubic-to-orthorhombic $\beta_1 \rightarrow \gamma_1'$ transformation. The strain-controlled loading program is specified by prescribing the macroscopic axial strain jointly with the condition that all macroscopic (but not local) stress components, except the macroscopic axial stress Σ, vanish. The macroscopic stress is thus expressed by $\mathbf{\Sigma} = \Sigma\,\mathbf{t} \otimes \mathbf{t}$, where $\mathbf{t}$ is a unit vector aligned with the tension axis. The needed material parameters (elastic constants and transformation strains) are provided in Appendix B.

The predicted overall stress–strain response in the stress-induced transformation is shown in Fig. 8.12 along with the evolution of microstructural parameters. The tension axis is specified by $\mathbf{t} = [0.925, 0.380, 0.]$, with respect to the cubic basis of austenite. The dashed lines in Fig. 8.12 correspond to the case of constant twin fraction λ, equal to that predicted by the crystallographic theory. Only the forward transformation from austenite to martensite is analyzed.

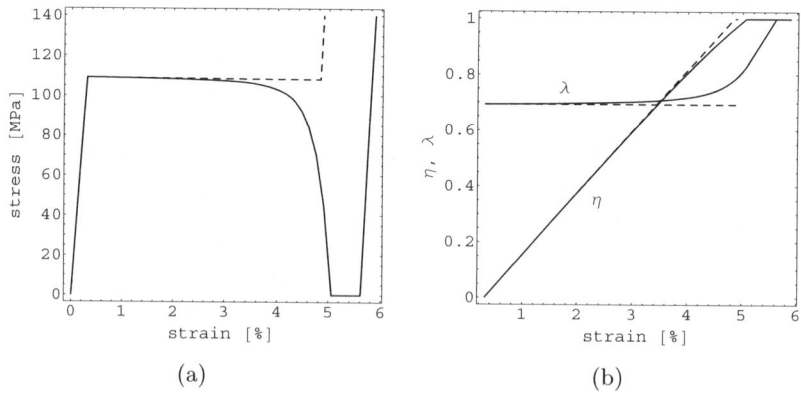

(a)   (b)

Fig. 8.12. Uniaxial tension of CuAlNi single crystal undergoing $\beta_1 \rightarrow \gamma_1'$ transformation: (a) stress–strain diagram and (b) evolution of twin fraction λ and volume fraction η of martensite (originally published in [141], copyright EDP Sciences).

It is seen in Fig. 8.12(b) that the volume fraction λ of one martensite variant within the plate grows substantially at the expense of the other variant near the end of transformation. This provides additional inelastic strain[9] compared to the case of fixed λ, but it also leads to a strongly negative slope of the macroscopic stress–strain diagram with a significant drop of the macroscopic stress, Fig. 8.12(a). At the instant when austenite disappears ($\eta = 1$), the overall stress falls to zero and further detwinning proceeds at zero stress until the less favorable variant disappears ($\lambda = 1$). This is followed by elastic loading of the remaining single variant of martensite. Qualitatively similar results have been obtained recently by Roytburd and Slutsker [124] who analyzed a cubic-to-tetragonal transformation, and assumed uniform and isotropic elastic properties of both phases.

To explain this somewhat surprising effect we note that, as austenite disappears, the average stress in a martensite plate tends to the overall stress $\boldsymbol{\Sigma}$. The driving force f^{JI} can thus be expressed in the limit $\eta = 1$ in terms of $\boldsymbol{\Sigma}$ as a sum of the leading term $\boldsymbol{\Sigma} \cdot (\varepsilon_I^t - \varepsilon_J^t) = \Sigma \mathbf{t} \cdot (\varepsilon_I^t - \varepsilon_J^t)\mathbf{t}$ and a quadratic correction term due to the mutual rotation of the elastic compliance tensors in martensite variants, in analogy to the formula (8.29). In the conditions met in the calculations above, the leading term with fixed $\mathbf{t} \cdot (\varepsilon_I^t - \varepsilon_J^t)\mathbf{t} \neq 0$ cannot be compensated by the correction term to produce $f^{JI} = 0$, unless $\Sigma = 0$ in the limit. Therefore, the overall stress must decrease to zero as austenite disappears; further variant rearrangement proceeds likewise at zero stress. The stress would not fall exactly to zero if a positive threshold value for f^{JI} was assumed, $f_c^{JI} > 0$, but the general behaviour would not change significantly for physically realistic twinning-related dissipation proportional to f_c^{JI}.

The behaviour illustrated in Fig. 8.12(a) by the solid line seems to be not in accord with typical stress–strain diagrams obtained from experimental tests on SMA specimens. However, it must be emphasized that the micromechanical model predicts the material behaviour under the assumption of development of a uniformly laminated microstructure within a material element. The transition from the material element scale to the scale of a single-crystal specimen or a grain in a polycrystal requires additional analysis. A qualitative analysis related to the latter case is provided in Sect. 8.4.3 below.

Hypothetical response of a single-crystal specimen associated with two possible patterns of localized transformation is sketched in Fig. 8.13. The expected instability of macroscopically uniform transformation may lead to localization of transformation zones. The negative tangent modulus for the material allows the transformation to proceed locally while the remaining part of the specimen undergoes elastic unloading. Moreover, the transformation may even be locally completed in a dynamic manner at fixed overall elongation Δl, corresponding to a local jump indicated schematically in Fig. 8.13 by a dashed

---

[9] The additional transformation strain due to detwinning may, in fact, be one of the reasons of the discrepancy between the observed transformation strains and the ones predicted under the assumption of fixed λ, cf. Fig. 8.8.

line. If such transformation takes place repeatedly in finite zones, then the resulting force–elongation $(P$–$\Delta l)$ diagram for the specimen may have the form sketched on the left-hand side of Fig. 8.13. Alternatively, smooth expansion of the fully transformed zone may lead to a diagram of the form sketched on the right-hand side of Fig. 8.13. Experimental stress–strain diagrams of both kinds have been reported in the literature, see for example Novák et al. [86, 87], Otsuka et al. [95], Zhang et al. [168]; see also Fig. 8.8. The instability phenomena are thus expected to play a crucial role in the microstructure evolution and in the overall behaviour of SMA specimens.

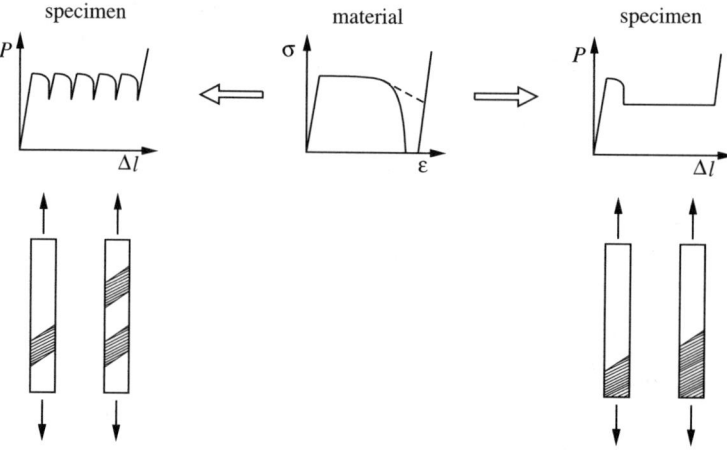

Fig. 8.13. Two hypothetical responses of a specimen with localized transformation zones (originally published in [141], copyright EDP Sciences).

8.4.3 Constrained Deformation Under Tension

In the case of uniaxial tension studied in Sect. 8.4.2, the only non-zero component of the overall stress tensor $\boldsymbol{\Sigma}$ is the axial component $\mathbf{t} \cdot \boldsymbol{\Sigma}\mathbf{t}$, equal to Σ'_{11} if the x'_1-axis of a Cartesian coordinate system is aligned with $\mathbf{t}$. This implies that the deformation is free in a sense that all the complementary components of the overall strain tensor $\mathbf{E}$ can take arbitrary values according to the macroscopic constitutive law. This is a very idealized situation, especially for a grain in a polycrystalline material, due to the constraints imposed by neighbouring grains, but also for an anisotropic tensile specimen with constrained grips. The effect of constrained deformation is therefore investigated in this section.

In order to study the effects of constrained deformation on transformation and detwinning, assume that, in addition to tensile loading, the following constraint is imposed on the overall strain

$$\mathbf{t} \cdot \mathbf{Es} = 0 \qquad \text{or} \qquad E'_{12} = 0, \tag{8.35}$$

where the unit vector $\mathbf{s}$ aligned with the x'_2-axis is perpendicular to the tensile loading direction $\mathbf{t}$, i.e. $\mathbf{t} \cdot \mathbf{s} = 0$. Clearly, the respective overall stress component $\mathbf{t} \cdot \mathbf{\Sigma s} = \Sigma'_{12}$ is not equal to zero in general. The constraint (8.35) approximately applies to a thin sheet-like specimen (with $\mathbf{s}$ and x'_2-axis lying within the sheet plane) subjected to tensile loading with the grips constrained laterally.

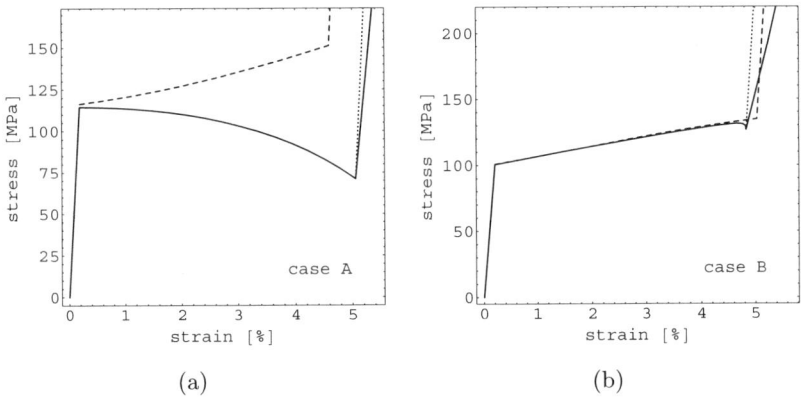

(a)   (b)

Fig. 8.14. Constrained tension of CuAlNi single crystal undergoing $\beta_1 \to \gamma'_1$ transformation (originally published in [141], copyright EDP Sciences).

The results obtained for two specific loading conditions are shown in Fig. 8.14. The tension axis $\mathbf{t} = [0.925, 0.380, 0.]$ is assumed as that in Sect. 8.4.2, and the constraint (8.35) is applied corresponding to two sheet orientations $\mathbf{s} = [-0.380, 0.925, 0.]$ (case A) or $\mathbf{s} = [-0.190, 0.463, 0.866]$ (case B), mutually rotated by 60 degrees. The resulting stress–strain diagrams are shown by the solid lines in Fig. 8.14, while the dashed lines correspond to the case of a fixed twin fraction λ. Comparison with Fig. 8.12(a) shows that the presence of the constraint (8.35) changes the material response significantly.

According to the assumption of the model, the twin fraction λ varies, so that the driving force on the twinning plane be equal to zero. In case A, the twin fraction of the most favorable martensite variant increases as the transformation proceeds from $\lambda = 0.711$ at $\eta = 0$ to $\lambda = 0.774$ at $\eta = 1$, cf. Fig. 8.15. Due to partial detwinning, the pseudoelastic strain at $\eta = 1$ is larger as compared to the case of fixed twin fraction, cf. Fig. 8.14(a). Further loading of the twinned martensite after the transformation is completed is accompanied by additional detwinning with the twin fraction reaching $\lambda = 0.8$ at the tensile stress of 201 MPa. For comparison, the purely elastic response of the twinned martensite with the twin fraction fixed at $\lambda = 0.774$ is marked in Fig. 8.14(a) by the dotted line. Clearly, detwinning provides additional strain

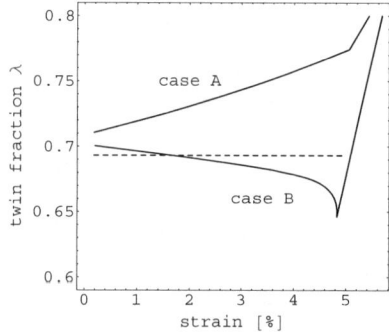

Fig. 8.15. Constrained tension of CuAlNi single crystal: variation of twin fraction λ (originally published in [141], copyright EDP Sciences).

Table 8.1. Effect of detwinning on the effective tangent stiffness modulus in constrained tension of CuAlNi crystal.

	case A	case B
martensite, detwinning	35.3 GPa	17.5 GPa
martensite, no detwinning	70.6 GPa	62.6 GPa
austenite	62.0 GPa	50.0 GPa

so that the response is more compliant. The values of the effective elastic and tangent stiffness moduli are given in Table 8.1.

A qualitatively different behaviour is predicted in case B. The twin fraction of the most favorable variant *decreases* during progressive transformation, cf. Fig. 8.15. This results in a hardening stress–strain response. Also, the pseudoelastic strain is somewhat smaller than in the case of constant twin fraction, cf. Fig. 8.14(b). However, once the transformation is completed, the elastic loading of martensite is associated with the growth of the favorable variant at the expense of the other one. The detwinning-related reduction of the effective tangent stiffness modulus is even more pronounced than in case A, cf. Table 8.1.

8.5 Summary and Conclusions

In the micromechanical modelling presented in this chapter, the stress-induced martensitic transformation in SMA single crystals has been assumed to proceed by evolution of laminated microstructures. In the simplest case of untwinned martensites, the microstructure is that of a rank-one laminate with the volume fraction of martensite as the only microstructural parameter that evolves during transformation. On the other hand, an evolving rank-two laminate is considered if, for instance, detwinning in internally twinned martensites is additionally accounted for. The corresponding micromechanical model, is based on a rate-independent transformation criterion, formulated in terms of the local thermodynamic driving force on the phase transformation front, and employs analytical micro-macro transition relations for laminated microstructures. Importantly, no extra kinetic equation is needed for the rate of the volume fraction of the product phase, since it is explicitly related to the macroscopic strain rate by the adopted transformation criterion; more precisely, by the consistency condition following from the transformation criterion. Elastic anisotropy of the phases is fully accounted for, and this is also a distinctive feature of the present model.

Macroscopic constitutive rate-equations have been derived for the case of an evolving rank-one laminate, i.e. for the situation when the product phase does not undergo any microstructural changes. These rate-equations have the form fully analogous to the equations of the classical rate-independent elastoplasticity. In particular, if the threshold thermodynamic driving force f_c is assumed to be constant, then a *softening* behaviour is predicted regardless of the elastic properties of the parent and product phases. In agreement with this theoretical result, the softening, though not substantial, is visible in the numerically simulated stress–strain response curves. Note that introducing some hardening into the transformation criterion, e.g. physically justified by material inhomogeneity, might lead to overall hardening response.

Practical implementation of the micromechanical model is based on the path-independent formulation outlined in Sect. 8.3.1. The computational scheme involves the explicit micro-macro transition relations for simple laminates, cf. Sect. 2.6 and Appendix A.4. As the elastic anisotropy of the phases is fully accounted for in the model, the resulting expressions, though available in an explicit form, become extremely complex. Accordingly, the symbolic code generation system *AceGen* [66] has been used to generate the respective numerical procedures.

The model is capable of reproducing important effects observed experimentally in single crystals subjected to uniaxial tension and compression, e.g. orientation dependence of the transformation stress, transformation strain, and elastic modulus. Furthermore, consistently with certain experimental observations, the area of the hysteresis loop of an isothermal load-displacement diagram is constant irrespective of the loading program executed.

In Sect. 8.3.4, a simple extension of the model has been proposed to account for non-isothermal effects. Two important effects are illustrated in a macroscopically adiabatic case: apparent hardening due to the latent heat of transformation and increase of the hysteresis width in the stress–strain diagram associated with heat conduction. The latter effect is observed if the forward and the reverse transformation proceed in adiabatic conditions, but cooling of the specimen is allowed after the forward transformation is completed and before the reverse transformation starts.

The general micromechanical modelling framework has also been applied for the case of internally twinned martensites, and the combined effect of the stress-induced transformation and detwinning has been investigated in Sect. 8.4. Evolution of the corresponding rank-two laminate has been examined assuming a negligible critical driving force on mobile interfaces between twin-related martensite variants. By the example of a CuAlNi single crystal undergoing the $\beta_1 \rightarrow \gamma_1'$ transformation, it has been shown that detwinning leads to a significant softening in the macroscopic response of a representative volume element. In the case of unconstrained deformation in uniaxial tension, the macroscopic stress can even drop to zero as the austenite disappears. The predictions of the model are not in accord with experimental observations. However, the model predicts the behaviour of a material point. Transition to the scale of a specimen or a grain in a polycrystal has not been attempted.

The effect of constrained deformation on the evolution of microstructure and on the macroscopic response has been illustrated by the example of tension with constrained shear. It has been shown that the difference in orientation of the shear constraint may lead to different detwinning effects, e.g. to the increase or decrease of the volume fraction of the favorable martensite variant in twinned martensite plates. On completing the austenite-to-martensite transformation, detwinning can still take place and influence the apparent stiffness of the martensite. Detwinning has been found to reduce significantly the effective tangent stiffness modulus of the twinned martensite. Hence, care is needed when interpreting apparent Young's modulus of martensite as representing its purely elastic stiffness.

The results of the present analysis indicate that the softening response is *always* expected at the material point, i.e. for a representative volume with a uniform laminated microstructure. The softening effect is rather small in untwinned martensites, but it can be substantial if detwinning is accounted for. Importantly, the experimental force-elongation diagrams of pseudoelastic SMA often exhibit features characteristic for unstable material response. The instability effects are thus expected to be a crucial element of modelling of pseudoelastic behaviour of shape memory alloys, and definitely deserve detailed studies. In fact, it can be shown that the macroscopically uniform transformation, associated with a fine, rank-one laminate at the micro-scale, is intrinsically unstable – see Petryk and Stupkiewicz [103] for preliminary results. Consequently, more complex transformation patterns may be expected to appear as a result of instability of uniform transformation. Furthermore, the stress–strain diagrams for a material element, as studied in this chapter, and for a specimen may be fundamentally distinct. In fact, the instability due to substantial softening associated with detwinning, as predicted by the model, cf. Sect. 8.4, is probably one of the reasons that the stress–strain response is qualitatively different depending on whether the product phase is untwinned or internally twinned – compare, for instance, the experimental curves in Figs. 8.7 and 8.8.

The small strain assumption, adopted in the present modelling, need not be a satisfactory approximation in view of transformation strains reaching, in some cases, 10% or even more. A finite strain formulation is thus desired, and a model employing the finite strain framework is currently under development, cf. Stupkiewicz and Petryk [139]. Note also, that the finite deformation framework is the correct one to consistently study the stability problems.

In the present applications of the model, microstructural parameters of martensitic plates, e.g. the habit plane orientation, are adopted from the classical crystallographic theory of martensite. This is, however, an approximation because the elastic strains, which are naturally not equal to zero in the stress-induced transformation, are neglected in that theory. The related effects are studied in the next chapter.

9

Formation of Stress-Induced Martensitic Plates

Abstract: An approach is developed for the prediction of transformation stresses and microstructures of stress-induced martensitic plates. Only the initial instant of transformation is considered, i.e. the formation of infinitely thin martensite plates within the austenite matrix. The microstructure is obtained as a solution of a constrained minimization problem for load multiplier. In the case of internally faulted martensites, the additional free energy associated with stacking faults in the martensite is accounted for, and a simple model relating this energy to the stacking fault energy and to the shear magnitude due to stacking faults is proposed. The approach is applied for CuZnAl single crystals undergoing stress-induced cubic-to-monoclinic transformation and the effects of stacking fault energy, loading direction, and temperature on the predicted microstructures are studied.

9.1 Minimization Problem for Load Multiplier

The crystallographic theory of martensite, outlined in Sect. 7.2, assumes stress-free conditions typical for transformations induced by changing temperature. Importantly, although Ball and James [8] developed the theory starting from elastic energy considerations within the framework of finite thermoelasticity, the actual problem is purely geometrical: compatibility conditions at zero stress are formulated, from which the microstructural parameters, e.g. the orientation of the austenite–martensite interface, are determined.

The compatible transformation strains and the corresponding microstructures predicted by the crystallographic theory are often adopted also in the case of stress-induced transformations, cf. Sect. 8.3. This is partially justified because the elastic deformations are typically much smaller than those resulting from the change of crystalline structure during transformation. However, in view of non-zero elastic strains naturally present in transformations under external stress, this is only an approximation.

On the other hand, the theory developed by Khachaturyan [59, 60] and Roytburd [117] predicts the optimal microstructural parameters by minimizing the elastic strain energy of a thin plate-like inclusion of the product phase formed within the parent phase matrix. Contrary to the crystallographic theory, this theory applies also for incompatible transformation strains. However, the related elastic strain effects are only accounted for within the small deformation framework. The results of this theory, mostly relevant to the present study, are summarized below. Assuming that the interfacial energy is sufficiently low, the inclusion of the product phase within the parent phase matrix has the form of a thin plate, except for special cases, e.g. isotropic elasticity with purely dilatational transformation strain. In the absence of external stresses, the orientation of the plate-like inclusion depends on the transformation strain, not necessarily compatible, and on elastic properties of the product phase. Under external stress, the microstructure is additionally affected by the difference of elastic moduli of parent and product phases. The practical applications of the theory for stress-induced transformations have so far been restricted to the case of the cubic-to-tetragonal transition, c.f. Roytburd and Pankova [121], Roytburd and Slutsker [122].

The second approach is further developed in this chapter with the aim to predict the microstructural parameters of stress-induced internally faulted martensitic plates. The general setting of the proposed approach is outlined in the present section. In the following sections the approach is specified for the case of internally faulted martensites by considering the additional free energy associated with stacking faults. It is assumed that, in addition to the usual variables, the free energy of martensite depends on the magnitude of shear induced by the stacking faults. Full account is also taken for the distinct elastic anisotropy of both phases. The resulting effects of the stacking fault energy, loading direction, and temperature on the microstructural parameters of internally faulted martensite plates are then studied for the cubic-to-monoclinic transformation in a CuZnAl shape memory alloy. The analysis is restricted to the initial instant of transformation when thin martensite plates are formed within the austenite matrix. The progressive transformation, with non-zero volume fractions of martensite, is not analyzed. The present chapter summarizes the results published in Stupkiewicz [131].

Consider thus the initiation of the stress-induced martensitic transformation, that is the situation when an infinitely thin plate of martensite (or many parallel martensite plates of total infinitesimal volume fraction) appears within the homogeneous austenite phase. Accordingly, we assume that initially, in the stress-free state, the crystal is in austenitic phase. Subsequently, a homogeneous external overall stress is applied. For simplicity we shall assume that the overall stress Σ varies proportionally,

$$\Sigma = p\Sigma^*, \tag{9.1}$$

where $p \geq 0$ is a load multiplier, not to be identified with the control parameter (e.g. strain) in the corresponding experiment, and Σ^* is a constant, prescribed reference stress (e.g. uniaxial tension or compression along a specified direction).

Using the micro-macro transition relations of Sect. 2.6, with the volume fraction of martensite set to zero, $\eta = 0$, the transformation condition $f - f_c = 0$, which follows from the phase transformation criterion (8.8), can be written in a general form

$$F(p, \mathcal{M}) = \hat{f}(p, \mathcal{M}) - f_c = 0, \tag{9.2}$$

where $f = \hat{f}(p, \mathcal{M})$ is the thermodynamic driving force at $\eta = 0$, and $\mathcal{M}$ denotes the set of all parameters determining the microstructure of a martensitic plate. This set includes the index I of the martensite variant, or the variant pair (I, J) in the case of an internally twinned plate, the orientation of the austenite–martensite interface specified by the normal vector $\mathbf{m}$, and possibly other parameters, e.g. the twin fraction λ. The microstructural parameters $\mathcal{M}$ and the load multiplier p, at which the transformation initiates, are unknown and have to be found for a given reference stress $\mathbf{\Sigma}^*$.

In order to determine the microstructure that would actually appear for the prescribed loading history, one can examine all the possible microstructures and select the one for which the transformation condition $F = f - f_c = 0$ is satisfied for the smallest load multiplier p. This can be written as a constrained minimization problem for the load multiplier p, namely

$$\boxed{\min_{\mathcal{M}} p \quad \text{subject to } F(p, \mathcal{M}) = 0.} \tag{9.3}$$

To be more specific, let us adopt the constitutive framework used in Sect. 8.3. Identifying the austenite with the '+' phase and the martensite with the '−' phase, the thermodynamic driving force (8.7) can be expressed as

$$f = -\Delta\phi_0 + \boldsymbol{\sigma}^+ \cdot \Delta\boldsymbol{\varepsilon}^t + \frac{1}{2}\boldsymbol{\sigma}^+ \cdot \Delta\mathbf{M}\boldsymbol{\sigma}^+ - \frac{1}{2}\Delta\boldsymbol{\sigma} \cdot \mathbf{M}^-\Delta\boldsymbol{\sigma}, \tag{9.4}$$

where $\boldsymbol{\sigma}^- = \boldsymbol{\sigma}^+ + \Delta\boldsymbol{\sigma}$ and $\Delta\phi_0 = \phi_0^- - \phi_0^+$. Using the interfacial relationships (2.33), the thermodynamic driving force can be expressed solely in terms of the stress $\boldsymbol{\sigma}^+$, namely

$$f = -\Delta\phi_0 + \boldsymbol{\sigma}^+ \cdot \Delta\boldsymbol{\varepsilon}^t + \frac{1}{2}\boldsymbol{\sigma}^+ \cdot \Delta\mathbf{M}\boldsymbol{\sigma}^+$$
$$-\frac{1}{2}(\Delta\mathbf{M}\boldsymbol{\sigma}^+ + \Delta\boldsymbol{\varepsilon}^t) \cdot \mathbf{S}^0(\Delta\mathbf{M}\boldsymbol{\sigma}^+ + \Delta\boldsymbol{\varepsilon}^t), \tag{9.5}$$

where the identity $(A.11)_2$ has also been used.

At the onset of transformation, the volume fraction of martensite is zero, $\eta = 0$, thus the stress within the austenite phase is equal to the overall one, $\boldsymbol{\sigma}^+ = \mathbf{\Sigma}$. Now, after substituting $\Delta^{am}\phi_0$, $\hat{\boldsymbol{\varepsilon}}_\alpha^t$, $\mathbf{M}_a$, and $\mathbf{M}_\alpha$ for $\Delta\phi_0$, $\Delta\boldsymbol{\varepsilon}^t$, $\mathbf{M}^+$, and $\mathbf{M}^-$, respectively, the thermodynamic driving force is given by

$$\hat{f}(p, \mathcal{M}) = -\Delta^{\mathrm{am}}\phi_0 + p\,\boldsymbol{\Sigma}^* \cdot \hat{\boldsymbol{\varepsilon}}_\alpha^{\mathrm{t}} + \frac{1}{2}p^2\boldsymbol{\Sigma}^* \cdot \Delta\mathbf{M}\,\boldsymbol{\Sigma}^*$$

$$-\frac{1}{2}\,(p\,\Delta\mathbf{M}\,\boldsymbol{\Sigma}^* + \hat{\boldsymbol{\varepsilon}}_\alpha^{\mathrm{t}}) \cdot \mathbf{S}^0(p\,\Delta\mathbf{M}\,\boldsymbol{\Sigma}^* + \hat{\boldsymbol{\varepsilon}}_\alpha^{\mathrm{t}}), \qquad (9.6)$$

where $\Delta\mathbf{M} = \mathbf{M}_\alpha - \mathbf{M}_{\mathrm{a}}$, $\mathbf{M}_{\mathrm{a}}$ is the elastic compliance of austenite, and $\mathbf{M}_\alpha$ and $\hat{\boldsymbol{\varepsilon}}_\alpha^{\mathrm{t}}$ are, respectively, the elastic compliance and the transformation strain of martensitic plate α. The dependence of f on the microstructural parameters $\mathcal{M}$ is also due to the dependence of $\mathbf{S}^0$ on the orientation of the austenite–martensite interface.[1] Finally, the transformation strain $\hat{\boldsymbol{\varepsilon}}_\alpha^{\mathrm{t}}$ and the effective elastic moduli tensor of the martensitic plate $\mathbf{M}_\alpha$ may depend on other microstructural parameters, e.g. on the twin fraction λ in the case of internally twinned martensite.

Remark 9.1. The transformation criterion $F = f - f_{\mathrm{c}} = 0$, which constitutes the constraint in the minimization problem (9.3), involves $\Delta\mathbf{M}$, the difference of elastic moduli tensors of both phases. In the case of distinct elasticity tensors, $\Delta\mathbf{M} \neq \mathbf{0}$, the transformation criterion is quadratic in the overall stress $\boldsymbol{\Sigma}$, i.e. quadratic in the load multiplier p, cf. (9.6). Then the solution of the minimization problem (9.3) depends on the actual value of the transformation stress, i.e. on the load multiplier p at which the transformation initiates. This dependence is implicit: the chemical energy $\Delta\phi_0^{\mathrm{am}}$ increases with temperature and so does the transformation stress, which is a part of the solution. Hence, at higher temperatures, the terms quadratic in $\boldsymbol{\Sigma}$ are more important, and the effects of distinct elastic properties of the phases are expected to be more pronounced.

Remark 9.2. Assume that $\Delta\mathbf{M} = \mathbf{0}$ and $\hat{\boldsymbol{\varepsilon}}_\alpha^{\mathrm{t}}$ is constant and depends only on the variant of martensite, so that the orientation of the austenite–martensite interface is the only unknown microstructural parameter. Then the constraint $F = 0$ is linear in p. Thus, for a fixed martensite variant, the interface normal $\mathbf{m}$ minimizing (9.3) does not depend on $\boldsymbol{\Sigma}$, and it can be found by maximizing the driving force f for a fixed load multiplier p, or equivalently by minimizing the last term in (9.6), i.e. the elastic strain energy of a thin plate-like inclusion $e^\infty = \frac{1}{2}\,\hat{\boldsymbol{\varepsilon}}_\alpha^{\mathrm{t}} \cdot \mathbf{S}^0\hat{\boldsymbol{\varepsilon}}_\alpha^{\mathrm{t}}$, cf. Khachaturyan [59, 60], Roytburd [117], Raniecki [110].

Remark 9.3. The chemical energy $\Delta^{\mathrm{am}}\phi_0$ and critical driving force f_{c} enter the minimization problem (9.3) only as a sum $\Delta^{\mathrm{am}}\phi_0 + f_{\mathrm{c}}$, through the constraint $F = f - f_{\mathrm{c}} = 0$. Therefore, the value of the sum $\Delta^{\mathrm{am}}\phi_0 + f_{\mathrm{c}}$ is sufficient to analyze the formation of martensite plates and the actual value of f_{c} is not relevant as long as the reverse transformation is not considered.

---

[1] The expression for $\mathbf{S}^0$ in (A.10) is given in the intrinsic coordinate system and thus $\mathbf{S}^0$ depends on $\mathbf{m}$.

9.2 Free Energy of Internally Faulted Martensites

9.2.1 Transformation Mechanism

The transformation mechanism in internally faulted martensites is briefly outlined in this section. The exposition below concerns the Cu-based shape memory alloys with the DO_3 type parent phase (e.g. CuZnAl, CuAlNi), however, it applies similarly also for the B2 type alloys (e.g. CuZn), cf. Otsuka et al. [91]. The atomic arrangement in the DO_3 type ordered structure is shown schematically in Fig. 9.1, cf. Otsuka et al. [91] and Hane [39]. The open and closed circles denote the positions that can be occupied by different atoms.

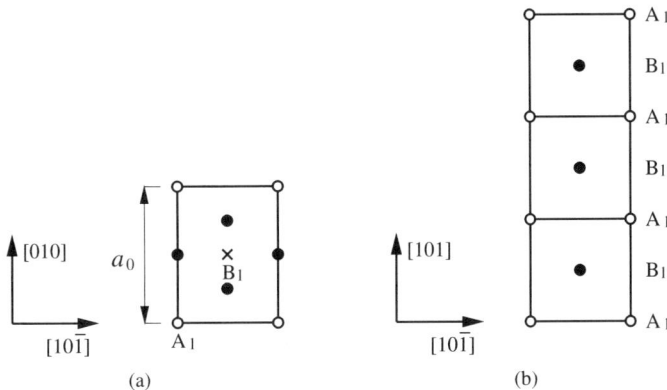

(a)   (b)

Fig. 9.1. The DO_3 type ordered structure: (a) atomic arrangement in the (101) basal plane; (b) stacking sequence viewed from the [010] direction. The open and closed circles denote the positions occupied by different atoms (reprinted from [131], Copyright 2004, with permission from Elsevier).

The martensitic transformation proceeds by the contraction along the [010] direction and by the expansion along the [101] and [10$\bar{1}$] directions. As a result the (101) basal planes become close-packed planes, cf. Otsuka et al. [91]. Also, the original stacking positions A_1 and B_1, cf. Fig. 9.1(a), are no longer stable, and the (101) basal planes move along the [10$\bar{1}$] direction to the nearest stable positions. The stable positions are placed at 0, $a/3$, and $2a/3$ and are denoted by A, B, C for the A_1 planes of the parent phase and by A', B', C' for the B_1 planes. These positions are indicated in Fig. 9.2(a). The actual stacking positions may deviate from the ideal $a/3$ and $2a/3$ positions; this is discussed later. The nearest stable positions for each basal plane require a shift by only $\pm a/6$ in the [10$\bar{1}$] direction with respect to the neighbouring basal plane. The ideal structure of martensite is obtained for a cyclic sequence of two shears to the left followed by one shear to the right as illustrated in Fig. 9.2(b). The monoclinic unit cell of martensite is indicated by a dashed line in Fig. 9.2(b).

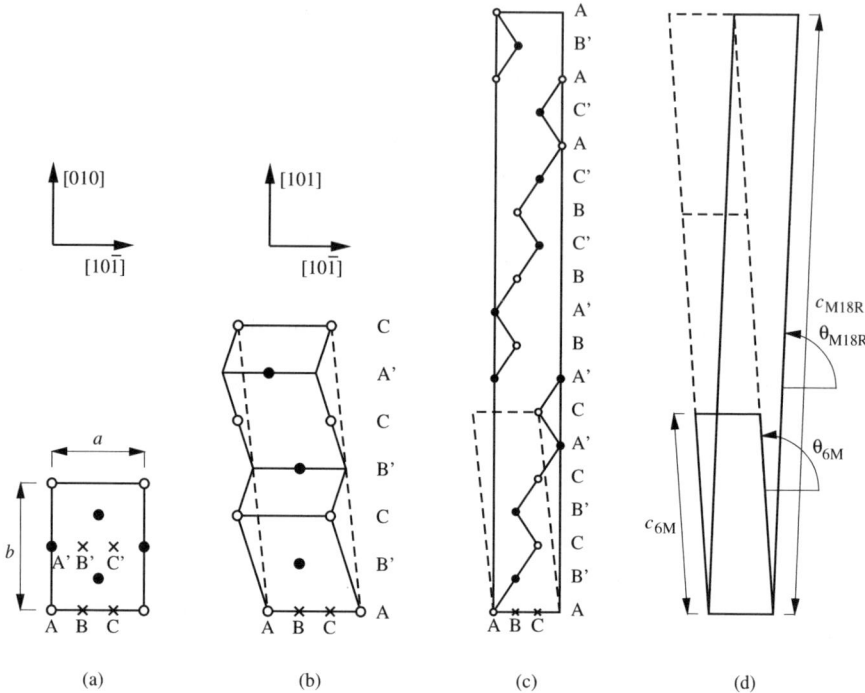

Fig. 9.2. The 6M and 18R unit cells of martensite: (a) the close-packed (101) basal plane; (b) the 6M unit cell (dashed lines); (c) the 18R unit cell; (d) the lattice parameters of 6M and M18R unit cells. Figures b, c and d show the view from the [010] direction, and crystallographic directions refer to the cubic basis of the parent phase (reprinted from [131], Copyright 2004, with permission from Elsevier).

It is called the *6M unit cell* as it involves six layers and the symmetry is monoclinic, cf. Otsuka et al. [91].

Historically, a different unit cell, called 18R (or M18R), was used to describe the same martensites, hence the common name *18R (M18R) martensites*. Indeed, in the case of ideal $a/3$, $2a/3$ stacking positions, an orthorhombic 18R unit cell can be constructed involving 18 basal planes with the AB'CB'CA'CA'BA'BC'BC'AC'AB'A stacking sequence. The 18R unit cell is shown in Fig. 9.2(c). When the stacking positions deviate from the ideal ones, the 18R unit cell becomes monoclinic with the monoclinic angle θ_{M18R} different from (but close to) 90°, hence the modified 18R unit cell, the *M18R unit cell*, was introduced. In fact, other unit cells can also be used to describe the crystalline lattice of these martensites. However, only the 6M unit cell corresponds to the actual transformation mechanism.

As already discussed in Sect. 7.2.3, direct austenite-single martensite variant interfaces are possible in the internally faulted martensites because of the additional shear due to stacking faults which provides compatibility at the austenite–martensite interface. A possible mechanism of stacking faulting in the internally faulted martensites is by perturbation of the ideal sequence of two shears to the left and one shear to the right, during the formation of the martensite plate. Such stacking faults are called *sequence faults*. Three types of sequence faults are shown in Fig. 9.3, cf. Andrade et al. [4]. These faults are introduced by a single violation of the ideal sequence, indicated by an arrow in Fig. 9.3, followed by the regular 2/1 shear sequence. Note that other types of sequence faults are also possible.

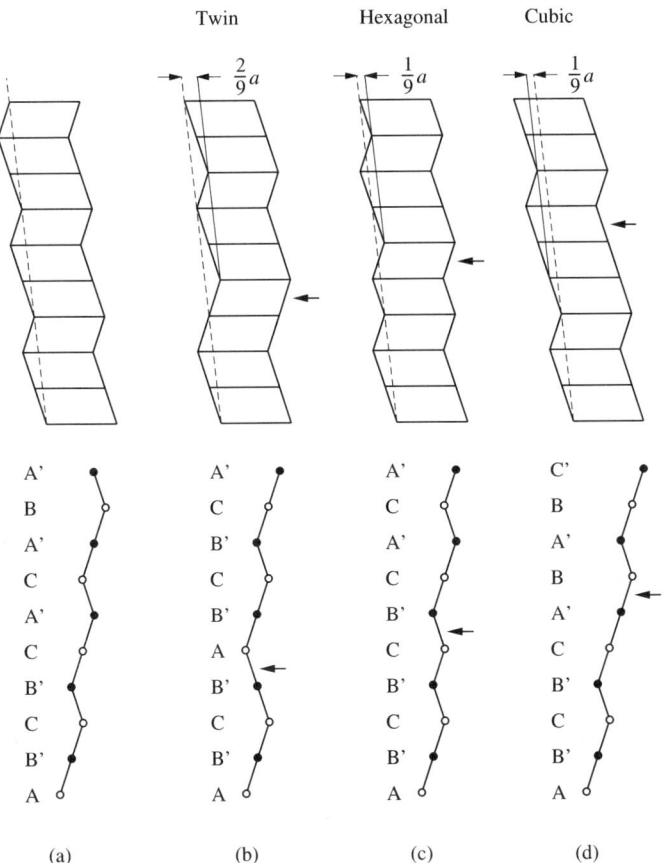

Fig. 9.3. The ideal structure (a) and sample sequence faults (b,c,d) in internally faulted 6M (M18R) martensites: shear sequences (top figures) and corresponding stacking sequences (bottom figures) (reprinted from [131], Copyright 2004, with permission from Elsevier).

9.2.2 Free Energy Due to Stacking Faults

In order to account for the effects of the energy associated with stacking faults, the free energy density of the I-th variant of martensite is assumed to have the form

$$\phi_I = \phi_0 + \Delta^{am}\hat{\phi}_0 + \frac{1}{2}(\varepsilon - \hat{\varepsilon}_I^t) \cdot \mathbf{L}_I(\varepsilon - \hat{\varepsilon}_I^t), \qquad (9.7)$$

where the term $\Delta^{am}\hat{\phi}_0$ is assumed to comprise two parts,

$$\Delta^{am}\hat{\phi}_0 = \Delta^{am}\phi_0 + \Delta\phi_0^{sf}. \qquad (9.8)$$

Here $\Delta^{am}\phi_0$ is the usual chemical energy, i.e. the difference of free energy densities between the martensite of ideal crystalline structure (i.e. without stacking faults) and the austenite, both in stress-free states. It is usually assumed that $\Delta^{am}\phi_0$ depends linearly on temperature, thus at a fixed temperature it is a constant.

The second term in (9.8), $\Delta\phi_0^{sf}$, is the additional free energy of martensite due to the presence of stacking faults. The stacking faults increase the internal energy u_m^0 of martensite in the reference state, cf. (8.5), while the entropy s_m^0 is not affected. A simple model relating this energy to the stacking fault energy (SFE) and to the shear magnitude k_{sf} is proposed below.

As already mentioned, there are many types of sequence faults that could possibly appear in 6M martensites. Each stacking fault type is characterized by the shear displacement, denote it by d, which is induced by a single fault. For example, $d = 2a/9$ for the twin-type fault, $d = a/9$ for the hexagonal-type fault, and $d = -a/9$ for the cubic-type fault, cf. Fig. 9.3. Moreover, each type of stacking fault can be characterized by a possibly different stacking fault energy Γ. The stacking fault energy Γ is the energy due to a single stacking fault per unit area of this fault.

Consider thus a $A \times B \times C$ cuboid of martensite, large enough to be treated as a representative volume, aligned with the 18R basis. Assume that the cuboid contains N stacking faults with (101) shear plane and [10$\bar{1}$] shear direction. Further, assume that all the faults are of the same type, so that their shear displacements d and stacking fault energies Γ are identical. This assumption is justified later. The overall shear magnitude is thus $k_{sf} = Nd/C$ and the additional free energy due to N stacking faults is $N\Gamma AB$. The volumetric density of this energy, i.e. the additional free energy density due to stacking faults, is thus

$$\boxed{\Delta\phi_0^{sf} = \frac{N\Gamma AB}{ABC} = \frac{\Gamma k_{sf}}{d} = \frac{\Gamma}{|d|}|k_{sf}|.} \qquad (9.9)$$

Equation (9.9) provides a relation between Γ, the *surface* density of the stacking fault energy, and the related *volumetric* density $\Delta\phi_0^{sf}$. The ratio k_{sf}/d is non-negative, so it has been replaced by $|k_{sf}|/|d|$ in (9.9), whereas the quantity $d/k_{sf} = C/N$ is the average distance between faulted planes.

Now, the question is, which of the many possible stacking fault types is actually formed during the transformation. One can expect that the one characterized by the lowest $\Gamma/|d|$ ratio, so that the additional shear required to ensure compatibility at the austenite–martensite interface requires as little energy as possible. The same line of argument supports the assumption that stacking faults of only one type appear during the transformation. Only in the situation when the ratios Γ'/d' and Γ''/d'' of two (or more) fault types are nearly equal, $\Gamma'/d' \approx \Gamma''/d'' \approx \Gamma/d$, both types could appear simultaneously. However, equation (9.9) holds also in that case, because the additional free energy density can then be written as $\Delta\phi_0^{sf} = \Gamma' k_{sf}'/d' + \Gamma'' k_{sf}''/d'' \approx (\Gamma/d)(k_{sf}' + k_{sf}'') = (\Gamma/d)k_{sf}$, where the total shear magnitude $k_{sf} = k_{sf}' + k_{sf}''$ is the sum of the shear magnitudes k_{sf}' and k_{sf}'' associated with each of the active fault types.

The ratio $\Gamma/|d| \geq 0$ of the energetically preferential stacking fault type (i.e. the one with the smallest $\Gamma/|d|$) is a material parameter and can be interpreted as the additional free energy density per unit shear magnitude. It will thus be called the *specific SFE density*. The specific SFE density may, in general, be different for $k_{sf} > 0$ and $k_{sf} < 0$, as positive and negative shears are generated by stacking faults of different types.

Comparing the free energy density of martensite, specified by (9.7)–(9.8), to that adopted in Sect. 8.2, cf. (8.4), we notice that stacking faulting in martensite is accounted for by the free energy term $\Delta\phi_0^{sf}$ in (9.8). Furthermore, in Sect. 8.2 the shear magnitude k_{sf} due to stacking faults is implicitly assumed constant, by adopting the microstructure predicted by the crystallographic theory, while here it will be determined from the minimization problem (9.3).

9.3 Internally Faulted Martensitic Plates

9.3.1 Minimization Problem

The general relations derived in Sect. 9.1 are now specified for the case of internally faulted martensites. The microstructure of the martensitic plate is now fully described by the variant index I, the shear magnitude k_{sf}, and the austenite–martensite interface normal $\mathbf{m}$. Using the free energy function (9.7), the driving force for the transformation from austenite to martensite variant I can be expressed as

$$f_I = \hat{f}_I(p, \mathbf{m}, k_{sf}) = -\Delta^{am}\hat{\phi}_0 + p\,\mathbf{\Sigma}^* \cdot \hat{\varepsilon}_I^t + \frac{1}{2}\,p^2 \mathbf{\Sigma}^* \cdot \Delta \mathbf{M}\mathbf{\Sigma}^*$$

$$-\frac{1}{2}\,(p\,\Delta \mathbf{M}\mathbf{\Sigma}^* + \hat{\varepsilon}_I^t) \cdot \mathbf{S}^0(p\,\Delta \mathbf{M}\mathbf{\Sigma}^* + \hat{\varepsilon}_I^t). \quad (9.10)$$

The driving force f_I depends on $\mathbf{m}$ through the operator $\mathbf{S}^0$ and on k_{sf} through $\Delta^{am}\hat{\phi}_0$, cf. (9.8)–(9.9), and through $\hat{\varepsilon}_I^t$, cf. (7.13). The minimization problem (9.3) takes now the form

$$\boxed{\min_{I,\mathbf{m},k_{\mathrm{sf}}} p \qquad \text{subject to } F_I(p,\mathbf{m},k_{\mathrm{sf}}) = 0,} \tag{9.11}$$

where $F_I(p,\mathbf{m},k_{\mathrm{sf}}) = \hat{f}_I(p,\mathbf{m},k_{\mathrm{sf}}) - f_{\mathrm{c}}$.

9.3.2 Solution Method

The constrained minimization problem (9.11) is a discrete-continuous optimization problem since I, the index of martensite variant, is one of the independent variables. However, since there are only 12 martensite variants, a continuous sub-problem

$$\min_{\mathbf{m},k_{\mathrm{sf}}} p \qquad \text{subject to } F_I(p,\mathbf{m},k_{\mathrm{sf}}) = 0 \text{ for given } I, \tag{9.12}$$

can be solved for each of the variants I and the solution with the smallest p can be chosen as the solution of problem (9.11). In fact, the number of variants for which problem (9.12) has to be solved can be significantly reduced by excluding the less favorable variants. For that purpose, the transformation stress predicted by the extension of the Schmid law can be used as an indicator.

The continuous constrained minimization problem (9.12) is solved in a standard way by introducing Lagrange multipliers and solving directly the optimality criteria. The corresponding Lagrangian L and the condition of stationary point of L (optimality criteria) are given by

$$L(\mathbf{x}) = p + \lambda F_I + \mu(\mathbf{m} \cdot \mathbf{m} - 1), \qquad \frac{\partial L}{\partial \mathbf{x}} = \mathbf{0}, \tag{9.13}$$

respectively, where λ and μ are Lagrange multipliers, and $\mathbf{x}$ is the vector of all unknowns, $\mathbf{x} = \{p,\mathbf{m},k_{\mathrm{sf}},\lambda,\mu\}$. The last term in $(9.13)_1$ is introduced to enforce $\mathbf{m}$ to be a unit vector.

The optimality criteria $(9.13)_2$ constitute a set of seven nonlinear equations which are solved using the iterative Newton method. The microstructural parameters predicted by the crystallographic theory, Sect. 7.2.4, are used as a starting point of the iterative procedure. Usually, this initial guess proves to be sufficiently close to the actual solution and no convergence problems are encountered. Special treatment is only required in some cases, mostly when the convergence is affected by the non-smoothness of the constraint $F_I = 0$ for k_{sf} close to zero. This is due to the term $|k_{\mathrm{sf}}|$ in (9.9).

Due to severe complexity of expressions involved in derivation of the gradient of the Lagrangian L, and its Hessian required for the Newton method, the symbolic code generation package *AceGen* [68] is used to automatically generate the respective numerical procedures.

9.3.3 Thermally Induced Transformation

Although this work is mainly concerned with the stress-induced transformation, the present approach can be formally applied to analyze the thermally induced transformation. Accordingly, the overall stress $\boldsymbol{\Sigma}$ is set to zero, and the temperature is assumed to vary slowly so that its uniformity is preserved. Similarly to the case of stress-induced transformation, the crystal is initially in austenitic state and the temperature is decreased until a thin plate of martensite is formed. Clearly, the formation of a thin plate-like inclusion of faulted martensite can only be studied in this way. Other, physically more relevant, transformation mechanisms (self-accommodating groups, twinning) are not considered, nor is the progressive transformation.

As the chemical energy $\Delta^{am}\phi_0$ increases linearly with temperature, cf. (8.5), in what follows, we shall treat $\Delta^{am}\phi_0$ as a temperature-like variable and use it as a control parameter instead of temperature. Putting $\boldsymbol{\Sigma} = \mathbf{0}$ and using (9.8)–(9.9), the transformation criterion $F_I = f_I - f_c = 0$ becomes

$$F_I(\Delta^{am}\phi_0, \mathbf{m}, k_{sf}) = -(\Delta^{am}\phi_0 + f_c) - \frac{\Gamma}{|d|}|k_{sf}| - \frac{1}{2}\hat{\boldsymbol{\varepsilon}}_I^t \cdot \mathbf{S}^0\hat{\boldsymbol{\varepsilon}}_I^t = 0. \quad (9.14)$$

We now look for the plate orientation $\mathbf{m}$ and for the shear magnitude k_{sf}, for which the transformation criterion $F_I = 0$ is satisfied at the maximum temperature, i.e. at maximum $\Delta\phi_0^{am}$. This leads to the following constrained minimization problem

$$\min_{\mathbf{m}, k_{sf}} -\Delta\phi_0^{am} \quad \text{subject to} \quad F_I(\Delta\phi_0^{am}, \mathbf{m}, k_{sf}) = 0. \quad (9.15)$$

Remark 9.4. The minimization problem (9.15) does not involve elastic properties of the austenite. Thus the resulting microstructure depends only on the transformation strain, elastic properties of the martensite, and specific SFE density.

Remark 9.5. As the overall stress is zero, $\boldsymbol{\Sigma} = \mathbf{0}$, there are no privileged directions, and thus all variants provide crystallographically equivalent solutions to problem (9.15).

Remark 9.6. Consider a compatible microstructure, i.e. assume that $\mathbf{m}$ and k_{sf} follow from the crystallographic theory, and thus the total transformation strain $\hat{\boldsymbol{\varepsilon}}_I^t$ satisfies the habit plane equation (7.14). Evaluating the optimality criteria $(9.13)_2$ for the compatible transformation strain, one can easily prove that $\partial L/\partial k_{sf} = -\text{sign}(k_{sf})\,\lambda\Gamma/|d| \neq 0$, where $L = -\Delta\phi_0^{am} + \lambda F_I + \mu(\mathbf{m} \cdot \mathbf{m} - 1)$ is the Lagrangian corresponding to the minimization problem (9.15). This means that the compatible microstructure is not optimal in the present sense. Only in the case $\Gamma/|d| = 0$, i.e. when there is no additional free energy due to stacking faults, the geometrically linear theory of Sect. 7.2.4 and the minimization problem (9.15) yield identical results.

9.3.4 A Property of the Minimization Problem (9.11)

The crystallographic theory provides two solutions of the habit plane equation (7.14) for each of the martensite variants. The two microstructures are crystallographically equivalent, i.e. one can be obtained from the other by applying a rotation belonging to the symmetry point group of the cubic austenite lattice, cf. Hane [39]. Additionally, in the *geometrically linear case*, the shape strain direction vector $\mathbf{b}/|\mathbf{b}|$ of one of the solutions is identical to the habit plane normal $\mathbf{m}$ of the other solution and conversely, so that $\mathbf{m}_1 = \mathbf{b}_2/|\mathbf{b}_2|$ and $\mathbf{m}_2 = \mathbf{b}_1/|\mathbf{b}_1|$, where $(\mathbf{m}_1, \mathbf{b}_1)$ and $(\mathbf{m}_2, \mathbf{b}_2)$ are the two solutions of the habit plane equation (7.14) for the same variant of martensite, cf. Bhattacharya [12], Stupkiewicz [131].

An interesting property of the minimization problem (9.11) has been observed in all cases analyzed numerically, although a theoretical proof is lacking. The problem (9.11) appears to have two solutions $(p^1, k_{\mathrm{sf}}^1, \mathbf{m}^1)$ and $(p^2, k_{\mathrm{sf}}^2, \mathbf{m}^2)$ with identical load multipliers $p^1 = p^2$ and shear magnitudes $k_{\mathrm{sf}}^1 = k_{\mathrm{sf}}^2$. However, in contrast to the crystallographic theory, the two microstructures are *not* crystallographically equivalent.

Furthermore, consider the jumps in the total strain corresponding to the two solutions, $\Delta\varepsilon^i = \frac{1}{2}(\mathbf{c}^i \otimes \mathbf{m}^i + \mathbf{m}^i \otimes \mathbf{c}^i)$, cf. the compatibility condition (2.27). It appears that the unit vector $\mathbf{c}^i/|\mathbf{c}^i|$ of one solution is identical to the interface normal of the other solution, so that $\mathbf{m}_1 = \mathbf{c}_2/|\mathbf{c}_2|$ and $\mathbf{m}_2 = \mathbf{c}_1/|\mathbf{c}_1|$. As a result, the total strain jump is *identical* for both solutions, $\Delta\varepsilon^1 = \Delta\varepsilon^2$.

9.4 Microstructures in CuZnAl Single Crystals

9.4.1 Material Parameters

In this section, the microstructures accompanying the formation of internally faulted martensitic plates in CuZnAl single crystals are studied using the approach proposed above. The availability of material parameters is rather restricted, and the CuZnAl alloy is, actually, the only one for which the elastic properties of single crystals of the austenite and the 6M (M18R) martensite could be found in the literature. The corresponding elastic constants are provided in Appendix B. The parameters characterizing the transformation strain are also provided in Appendix B.

Three additional parameters are required to fully characterize the material: the chemical energy $\Delta^{\mathrm{am}}\phi_0$, the critical driving force f_c, and the specific SFE density $\Gamma/|d|$. As only the forward transformation is considered here, without loosing generality, $f_c = 0$ is assumed in the present simulations, so that $\Delta^{\mathrm{am}}\phi_0$ has actually the meaning of $\Delta^{\mathrm{am}}\phi_0 + f_c$, see Remark 9.3. In the examples below, a range of values of the chemical energy is used, $\Delta^{\mathrm{am}}\phi_0 = 0\div20\,\mathrm{MJ/m}^3$, which covers the range of realistic transformation stresses. Note that, since $\Delta^{\mathrm{am}}\phi_0$ depends on temperature, the adopted range of $\Delta^{\mathrm{am}}\phi_0$ corresponds to some range of temperatures which, however, is not specified here.

Determination of the specific SFE density $\Gamma/|d|$ would be possible if the stacking fault energies Γ of the possible sequence faults were known, cf.

Sect. 9.2.2. Since these parameters are not available, a range of values of the specific SFE density from $\Gamma/|d| = 0$ up to $\Gamma/|d| = 300\,\mathrm{MJ/m^3}$ is investigated in the examples. Assuming that $|d| = a/9$ (as for the cubic- and hexagonal-type faults), the value of $\Gamma/|d| = 300\,\mathrm{MJ/m^3}$ corresponds to the stacking fault energy $\Gamma = 15.1\,\mathrm{mJ/m^2}$.

9.4.2 Thermally Induced Transformation

The solution of the minimization problem (9.15), i.e. the chemical energy $\Delta^{\mathrm{am}}\phi_0$ at the initial instant of transformation induced thermally at zero stress and the corresponding shear magnitude k_{sf} are presented in Fig. 9.4 as a function of the specific SFE density $\Gamma/|d|$.

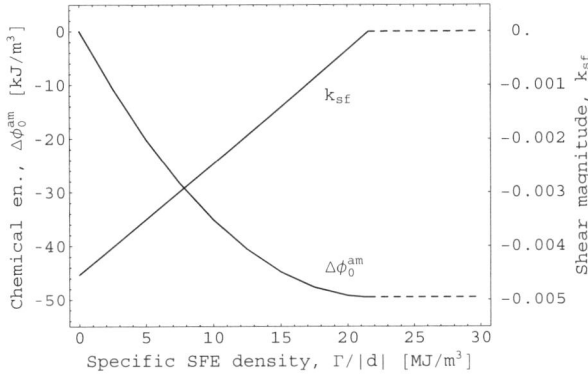

Fig. 9.4. Chemical energy $\Delta^{\mathrm{am}}\phi_0$ and shear magnitude k_{sf} as a function of the specific SFE density $\Gamma/|d|$ at the initiation of the thermally induced martensitic transformation.

As expected, the solution of the minimization problem (9.15) for $\Gamma/|d| = 0$ and the prediction of the geometrically linear crystallographic theory, cf. Table B.2, are identical. Furthermore, for $\Gamma/|d| = 0$, the chemical energy minimizing (9.15) is $\Delta^{\mathrm{am}}\phi_0 = 0$. Note, however, that $f_{\mathrm{c}} = 0$ has been assumed. Thus, for non-zero critical driving force ($f_{\mathrm{c}} > 0$), over-cooling ($\Delta^{\mathrm{am}}\phi_0 < 0$) is required to initiate the transformation.

With increasing specific SFE density $\Gamma/|d|$, the shear magnitude k_{sf} decreases in absolute value, Fig. 9.4(b), so that the total transformation strain $\hat{\varepsilon}^t$ is no longer compatible, cf. Remark 9.6. This is accompanied by the decrease of the chemical energy $\Delta^{\mathrm{am}}\phi_0$ (i.e. over-cooling) required to overcome this incompatibility, cf. Fig. 9.4(a). At a critical value $\Gamma/|d| \approx 21.5\,\mathrm{MJ/m^3}$, the shear magnitude k_{sf} reaches zero, and the chemical energy drops to $\Delta^{\mathrm{am}}\phi_0 = -49.5\,\mathrm{kJ/m^3}$. Assuming that $\Delta s^* = 0.156\,\mathrm{MJ/m^3}$ as in the case of CuAlNi alloy in Sect. 8.3.4, this corresponds to the overcooling of only 0.3 K. With the further increase of $\Gamma/|d|$, the solution is insensitive to $\Gamma/|d|$, as indicated by the dashed lines in Fig. 9.4.

The predicted orientation of the austenite–martensite interface also depends on $\Gamma/|d|$. However, the maximum deviation of the normal vector $\mathbf{m}$ (corresponding to the high values of $\Gamma/|d|$ and $k_{sf} = 0$) from the one predicted by the crystallographic theory is only about 0.6 degree.

9.4.3 Uniaxial Tension and Compression

The stress tensor corresponding to uniaxial tension and compression can be written as $\boldsymbol{\Sigma} = \Sigma\,\mathbf{t}\otimes\mathbf{t}$, where $\mathbf{t}$ is the unit vector defining the loading direction, and Σ is the uniaxial stress. The load multiplier is thus defined as $p = \Sigma > 0$ for tension and $p = -\Sigma > 0$ for compression. The unit vectors $\mathbf{t}$, corresponding to six loading directions used in the present study, are shown in Fig. 9.5. These directions are chosen arbitrarily in a way to cover different areas of the unit stereographic triangle. The twelve loading cases (tension or compression in six directions) are denoted by a character specifying the direction followed by 't' for tension or 'c' for compression. Thus, for example, A-c denotes compression in direction A.

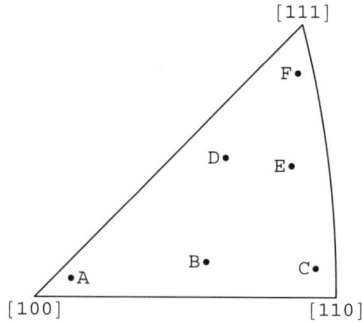

Fig. 9.5. Load axis orientations relative to the cubic basis of austenite.

The solution of the minimization problem (9.11) provides the variant I, the shear magnitude k_{sf}, the habit plane normal $\mathbf{m}$, and the uniaxial transformation stress Σ. For fixed elastic and transformation strain parameters, as specified in Appendix B, the solution depends on the chemical energy $\Delta^{am}\phi_0$ (i.e. indirectly on temperature), on the specific SFE density $\Gamma/|d|$, and on the loading direction. The influence of these three factors on the predicted microstructures is studied below.

The preferred martensite variant I appears to be fully determined by the sense of loading (this does not necessarily hold for loading directions other than the analyzed directions A,...,F). The variant $I = 9$ is the preferred variant in tension (A-t,...,F-t), and $I = 4$ is the preferred variant in compression (A-c,...,F-c).

In Figure 9.6, the shear magnitude k_{sf} is shown as a function of the specific SFE density $\Gamma/|d|$ for a representative loading case and for different values of the chemical energy $\Delta^{am}\phi_0$. Qualitatively, the dependence of k_{sf} on $\Gamma/|d|$ is similar to that obtained for the thermally induced transformation. With increasing $\Gamma/|d|$, the shear magnitude k_{sf} decreases in absolute value, and a critical value of the specific SFE density $\Gamma/|d|$ exists for which martensite plates with no stacking faults ($k_{sf} = 0$) are predicted. The critical specific SFE density increases with increasing $\Delta^{am}\phi_0$. The shear magnitude k_{sf} depends strongly on the loading direction, and so does the critical specific SFE density. This is clearly seen in Fig. 9.7, where k_{sf} is shown as a function of $\Gamma/|d|$ for $\Delta^{am}\phi_0 = 10\,\mathrm{MJ/m^3}$ and for all loading directions in tension.

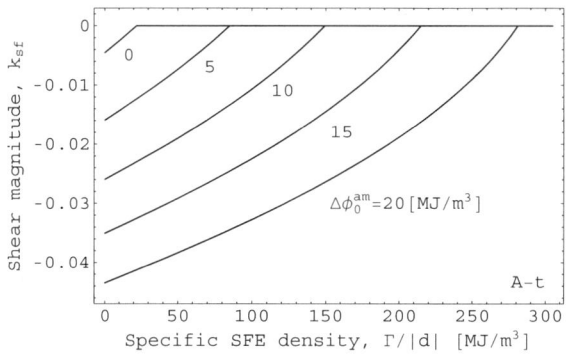

Fig. 9.6. Shear magnitude k_{sf} as a function of the specific SFE density $\Gamma/|d|$ for tension in direction A (reprinted from [131], Copyright 2004, with permission from Elsevier).

The dependence of the solution of the minimization problem (9.11) on the specific SFE density $\Gamma/|d|$ is non-smooth at the critical specific SFE density, i.e. when k_{sf} approaches zero as $\Gamma/|d|$ increases, cf. Figs. 9.6 and 9.7. This is because the constraint $F_I = f_I - f_c = 0$, which involves $|k_{sf}|$, is non-smooth at $k_{sf} = 0$.

The transformation stress Σ as a function of $\Gamma/|d|$ is shown in Fig. 9.8 for tension in direction A. For $\Gamma/|d|$ smaller than the critical value, the transformation stress increases with increasing $\Gamma/|d|$, for higher values the transformation stress is constant. The dashed line labelled by '$k_{sf} = 0$' in Fig. 9.8 indicates the critical specific SFE densities and the corresponding transformation stresses at different values of the chemical energy $\Delta^{am}\phi_0$. The transformation stresses predicted for $\Delta^{am}\phi_0 = 10\,\mathrm{MJ/m^3}$ are given in Table 9.1. For each loading case, the smallest and the largest transformation stress, corresponding to

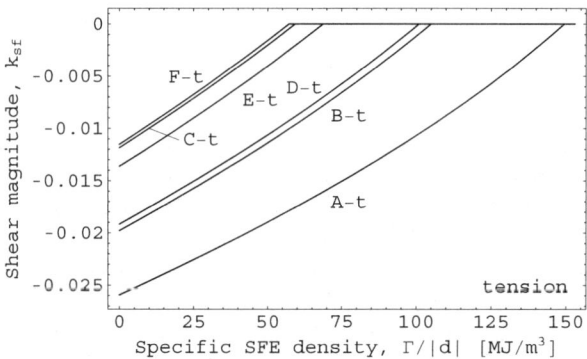

Fig. 9.7. Shear magnitude k_{sf} as a function of the specific SFE density $\Gamma/|d|$ in tension for $\Delta^{\mathrm{am}}\phi_0 = 10\,\mathrm{MJ/m^3}$ (reprinted from [131], Copyright 2004, with permission from Elsevier).

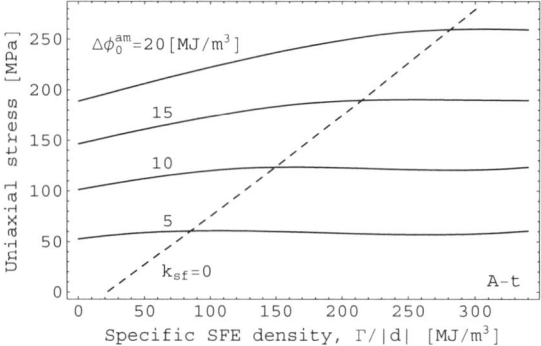

Fig. 9.8. Uniaxial transformation stress Σ as a function of the specific SFE density $\Gamma/|d|$ for tension in direction A (reprinted from [131], Copyright 2004, with permission from Elsevier).

the limit cases $\Gamma/|d| = 0$ and $k_{\mathrm{sf}} = 0$, respectively, are included in Table 9.1 along with their relative difference.

Finally, Fig. 9.9 illustrates the effect of the specific SFE density $\Gamma/|d|$ on the orientation of the habit plane: the habit plane normal vectors are shown in stereographic projection for $\Delta^{\mathrm{am}}\phi_0 = 20\,\mathrm{MJ/m^3}$ and for two representative loading cases (A-t and F-t). As the specific SFE density $\Gamma/|d|$ increases from zero to its critical value, the habit plane normal evolves between two limit cases denoted by $\Gamma/|d| = 0$ and $k_{\mathrm{sf}} = 0$ in Fig. 9.9. The habit plane normal vector predicted by the crystallographic theory, marked with a triangle and a 'CT' label, is also included in Fig. 9.9.

As already discussed in Sect. 9.3.4, two solutions of the minimization problem (9.11) are obtained, for which the martensite variant I, the shear magnitude k_{sf}, and the transformation stress Σ are identical. However, the habit

Table 9.1. Predicted values of uniaxial transformation stress corresponding to different loading cases ($\Delta^{\mathrm{am}}\phi_0 = 10\,\mathrm{MJ/m^3}$).

Loading	Σ (MPa)		Relative
case	$\Gamma/|d| = 0$	$k_{\mathrm{sf}} = 0$	difference
A-t	101.3	123.5	21.9%
B-t	105.0	116.7	11.1%
C-t	127.1	131.6	3.6%
D-t	138.3	152.6	10.3%
E-t	150.5	157.7	4.8%
F-t	268.2	277.2	3.4%
A-c	-107.7	-111.5	3.5%
B-c	-133.7	-141.6	5.9%
C-c	-194.6	-226.6	16.4%
D-c	-170.9	-178.4	4.4%
E-c	-202.1	-220.1	8.9%
F-c	-338.7	-351.8	3.8%

plane normal vectors $\mathbf{m}_1$ and $\mathbf{m}_2$ are not crystallographically equivalent. In order to make the comparison possible, the vectors corresponding to both solutions are shown in the same unit stereographic triangle (in practice, both vectors are approximately perpendicular). The difference between the two solutions obtained for tension in direction A is hardly visible in Fig. 9.9(a). The difference is much more pronounced when tension in direction F is considered, cf. Fig. 9.9(b).

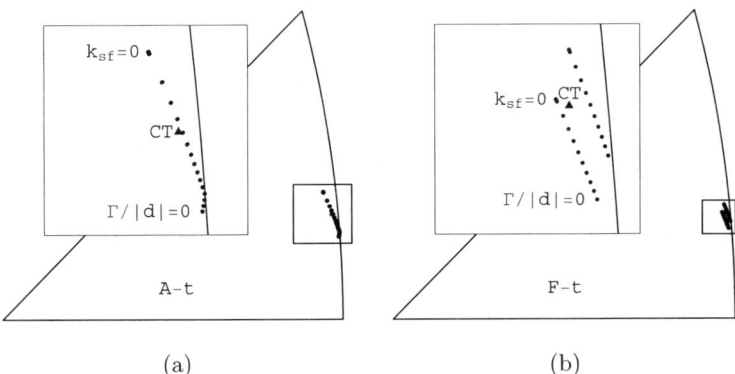

(a)   (b)

Fig. 9.9. The habit plane normal vectors $\mathbf{m}$, corresponding to different values of $\Gamma/|d|$, shown in the unit stereographic triangle (reprinted from [131], Copyright 2004, with permission from Elsevier).

The habit plane orientations predicted for the two limit cases $\Gamma/|d| = 0$ and $k_{\mathrm{sf}} = 0$ are shown in Fig. 9.10. All the loading cases (tension and compression in directions A,...,F) are included in Fig. 9.10. The predicted habit plane orientations form two separate groups corresponding to $\Gamma/|d| = 0$ and to $k_{\rm sf} = 0$. The maximum deviation from the mean orientation within the group depends on $\Delta^{\rm am}\phi_0$. At $\Delta^{\rm am}\phi_0 = 20\,{\rm MJ/m^3}$, the maximum deviation is 1.3 degree for $\Gamma/|d| = 0$ and 1.8 degree for $k_{\rm sf} = 0$. At $\Delta^{\rm am}\phi_0 = 10\,{\rm MJ/m^3}$, the deviation is about two times smaller.

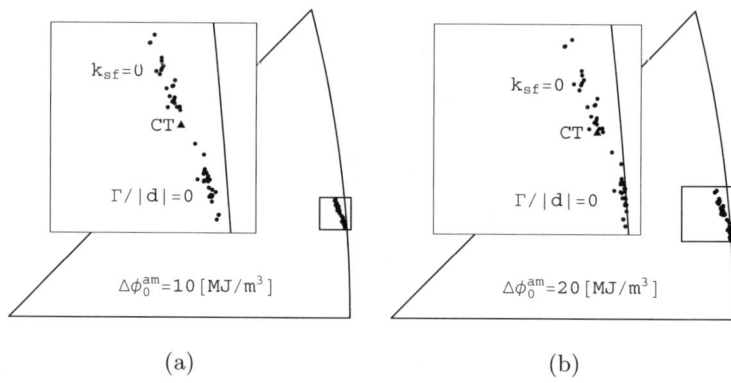

(a)   (b)

Fig. 9.10. Habit plane normal vectors **m** predicted for all loading cases and for $\Gamma/|d| = 0$ or for $k_{\rm sf} = 0$, shown in the unit stereographic triangle (reprinted from [131], Copyright 2004, with permission from Elsevier).

9.5 Discussion

An approach has been developed for prediction of transformation stresses and microstructures of stress-induced internally faulted martensitic plates at the initial instant of transformation. The problem has been formulated as a minimization problem for the load multiplier with the transformation condition, expressed in terms of the thermodynamic driving force on the phase transformation front, imposed as a constraint. The general setting of the problem has been provided, which has then been specified for the internally faulted martensites. The latter case has been further studied in the numerical examples. The additional free energy associated with stacking faults in the martensite has been accounted for, and a simple model relating this energy to the stacking fault energy and to the shear magnitude due to stacking faults has been proposed.

   The predicted microstructures are, in general, different from the one following from the crystallographic theory, both in terms of the habit plane orientation and of the shear magnitude due to stacking faults. Moreover, the microstructures depend on the stress state, i.e. on the orientation of the tension or compression axis with respect to the austenite lattice and on the stress magnitude. The latter dependence is implicit since the transformation stress, as a part of the solution, depends on the chemical energy $\Delta\phi_0^{am}$, which, in turn, depends on the temperature.

The microstructure depends on the stress state only if the elastic moduli tensors of austenite and martensite are different. This difference, however, is always expected in view of different crystallographic symmetry and related different elastic anisotropy of both phases. The higher the temperature, the higher the transformation stress and the more pronounced the elastic mismatch effects, e.g. the scatter of habit plane orientations. The examples cover uniaxial tension and compression only, however, both the method and the results are valid for general stress states, e.g. those induced in differently oriented grains of a loaded polycrystalline material.

The additional free energy associated with stacking faults is an important factor affecting the predicted microstructures, and the specific SFE density $\Gamma/|d|$ has been shown to be the influential parameter. The specific SFE density is a material parameter having, in general, two distinct values corresponding to $k_{sf} > 0$ and to $k_{sf} < 0$. Determination of the specific SFE density requires that the sequence-fault type, characterized by the lowest $\Gamma/|d|$ ratio, is chosen from the many possible types. Unfortunately, the actual values of stacking fault energies Γ of different sequence-type stacking faults are not available, so a parametric study could only be performed.

Based on the results presented in Sect. 9.4, three situations can be predicted. For low stacking fault energy (more precisely for low specific SFE density $\Gamma/|d|$), all martensitic plates, regardless of the stress state, would be internally faulted. The amount of shearing due to stacking faults, closely related to the average distance between the faulted planes, would depend on the stress state. On the other hand, if the stacking fault energy was sufficiently high, all martensitic plates, regardless of the stress state, would be formed without stacking faults. Finally, in an intermediate situation, depending on the stress state, some martensite plates would be internally faulted and some would be not. The above division is relative and changes with temperature.

As the specific SFE density $\Gamma/|d|$ increases, the solution $(p, k_{sf}, \mathbf{m})$ of the minimization problem monotonically evolves between two limit cases $\Gamma/|d| = 0$ and $k_{sf} = 0$. The load multiplier p, at which the transformation initiates, is greater for $k_{sf} = 0$ than for $\Gamma/|d| = 0$. This is expected, since the condition $k_{sf} = 0$ can, in fact, be interpreted as an additional constraint on the minimization problem (9.11). Interestingly, in both limit cases, the additional energy due to stacking faults vanishes, $\Delta\phi_0^{sf} = (\Gamma/|d|)|k_{sf}| = 0$.

Possible experimental verification of the model would require exact determination of microstructural parameters of martensite plates. The dependence of the habit plane orientation on the stress state (orientation and magnitude) may be difficult to observe experimentally, since the predicted variation of habit plane orientation with changing loading direction is of the order of two-three degrees which is, probably, close to the measurement error. Note, however, that, for differently oriented single crystal specimens, Horikawa et al. [48] reported different habit plane orientations, and the maximum deviation of the measured orientation from the one predicted by the crystallographic theory was about 3 degrees. Scatter of habit plane orientations has also been observed in polycrystalline materials undergoing stress-induced martensitic transformations, cf. Roytburd and Pankova [121], Zimmermann and Humbert [170]. Another related effect that could be approached experimentally is the dependence of the density of stacking faults on temperature and loading direction.

10
Final Remarks

Several aspects of micromechanics of contact and interphase layers have been addressed in this monograph. All the specific problems and applications are concerned with interfaces, interface layers, or materials with propagating interfaces. The reported studies have thus the *interfaces* in common.

Micromechanics provides a link between the structure and the properties at different scales of observation. It is *advantageous* because the macroscopic behaviour is analyzed by considering the microscopic interaction mechanisms which are usually better understood. Moreover, parameters characterizing the material at the micro-scale have a clear physical interpretation, and often can be measured directly.

It has been shown in Sect. 8.3 that several features of the complex pseudoelastic behaviour of SMA single crystals can be properly described by the proposed model employing the micromechanical approach combined with few, rather natural assumptions. The involved material parameters (elastic constants, transformation strains, chemical energy), indeed, have a clear physical meaning and can be measured directly – at least in principle, as there are severe difficulties in finding the elastic properties of single crystals in martensitic state. In the present model, only one parameter, the critical driving force f_c, has a more phenomenological character, however, as it is closely related to the intrinsic dissipation, it can be determined from the width of the hysteresis loop on an isothermal stress–strain diagram. Note that, as illustrated in Sect. 8.3.4, heat conduction in non-isothermal conditions may influence the apparent width of the hysteresis loop.

Similarly, successful prediction of the elasto-plastic normal contact compliance has been reported in Sect. 6.4. Surface roughness topography and local material properties are the only microstructural properties required as the input for the finite element model of the contact boundary layer. Measurement of the three-dimensional surface topography is a standard procedure, also the elastic and plastic properties of the bulk material can be measured easily. However, determination of the distribution of plastic properties within the surface layer is much more difficult. This has not been attempted, and a simplifying assumption of a constant yield stress has been adopted leading to satisfactory results.

However, the above example indicates one of the *limitations* of the micromechanical approach: its success may by affected by the lack of detailed and accurate characterization of the microstructure, micro-mechanisms, or local material properties. Let us also mention another limitation of the micromechanical modelling: analysis of phenomena occurring at several scales may be necessary in order to capture all essential deformation mechanisms. Such multi-scale analysis might be very expensive in terms of computational cost, it might also be very demanding regarding the modelling assumptions (microstructure, local material behaviour) at each of the analyzed scales. Thus, an alternative combined approach, i.e. a phenomenological description based on sound micromechanical reasoning, should be developed in parallel.

This book illustrates also the central role of the *compatibility conditions* in the micromechanical analysis of interfaces and interface layers. In fact, these conditions appear in some form in each chapter, usually as an essential element of the modelling. The local compatibility conditions at a bonded interface stem from the assumptions of displacement continuity and mechanical equilibrium, cf. Sect. 2.4. In the case of laminates and thin layers, the stresses and strains are piecewise homogeneous, so that the compatibility conditions trivially hold also for the strains and stresses away from the interfaces. Finally, in the boundary layers, the strains and stresses are inhomogeneous, but the compatibility conditions hold then for the respective averages, cf. Sect. 5.2.

In Sect. 6.4, the finite element model of a contact boundary layer has been developed for a real three-dimensional roughness topography. This resulted in a moderately large-scale simulation with more than 50,000 unknowns, although the discretization of the contact surface was not particularly fine (55×55 nodes). Clearly, the size of the numerical problem is a limitation for the refinement of discretization. Nevertheless, direct micromechanical simulations of rough contact interactions are expected to be more and more frequently used in engineering practice. This is mostly due to continuously increasing available computing power, so that more and more realistic problems can be approached at a reasonable computational cost.

A

Micro-Macro Transition Relations for Simple Laminates

A.1 Matrix Notation

In this appendix, explicit expressions are provided for the interfacial operators, concentration matrices, and effective elastic matrices introduced in the tensor notation in Chap. 2. The derivation methodology is described in Chap. 2, therefore the details are omitted here.

The matrix notation is used throughout this appendix.[1] The advantage is also taken of the interior–exterior decomposition in the intrinsic coordinate system associated with the interface. This notation is a convenient tool for derivation of explicit expressions for the related operators. Also, thanks to the interior–exterior decomposition, the form of obtained expressions is transparent and allows easy interpretation of some properties.

In the Kelvin matrix notation, the Cauchy stress tensor and the infinitesimal strain tensor (both symmetric) are represented by respective vectors in a six-dimensional space, see for example Cowin and Mehrabadi [24] and Pedersen [99],

$$
\begin{aligned}
\boldsymbol{\sigma} &= \{\boldsymbol{\sigma}_N, \boldsymbol{\sigma}_S\} = \{\{\sigma_{11}, \sigma_{22}, \sigma_{33}\}, \sqrt{2}\{\sigma_{12}, \sigma_{13}, \sigma_{23}\}\}, \\
\boldsymbol{\varepsilon} &= \{\boldsymbol{\varepsilon}_N, \boldsymbol{\varepsilon}_S\} = \{\{\varepsilon_{11}, \varepsilon_{22}, \varepsilon_{33}\}, \sqrt{2}\{\varepsilon_{12}, \varepsilon_{13}, \varepsilon_{23}\}\},
\end{aligned}
\tag{A.1}
$$

and the fourth-rank tensors, e.g. the elasticity tensors, are represented by 6×6 matrices. For instance, the elastic stiffness matrix $\mathbf{L}$ has the form

$$
\mathbf{L} = \begin{bmatrix} \mathbf{L}_{NN} & \mathbf{L}_{NS} \\ \mathbf{L}_{NS}^T & \mathbf{L}_{SS} \end{bmatrix},
\tag{A.2}
$$

where the 3×3 submatrices are given by

---

[1] For this reason, some notation conventions used in this appendix are not fully consistent with the notation used in the remainder of this book. The deviations in the notation are, however, self-explanatory and detailed definitions are not provided.

$$\mathbf{L}_{NN} = \begin{bmatrix} L_{1111} & L_{1122} & L_{1133} \\ & L_{2222} & L_{2233} \\ \text{sym.} & & L_{3333} \end{bmatrix}, \quad \mathbf{L}_{SS} = 2 \begin{bmatrix} L_{1212} & L_{1213} & L_{1223} \\ & L_{1313} & L_{1323} \\ \text{sym.} & & L_{2323} \end{bmatrix},$$

$$\mathbf{L}_{NS} = \sqrt{2} \begin{bmatrix} L_{1112} & L_{1113} & L_{1123} \\ L_{2212} & L_{2213} & L_{2223} \\ L_{3312} & L_{3313} & L_{3323} \end{bmatrix},$$

and L_{ijkl} are the components of the fourth-rank elastic stiffness tensor, such that $\sigma_{ij} = L_{ijkl}\varepsilon_{kl}$. The constitutive relations,

$$\boldsymbol{\sigma} = \mathbf{L}(\boldsymbol{\varepsilon} - \boldsymbol{\varepsilon}^t), \qquad \boldsymbol{\varepsilon} = \mathbf{M}\boldsymbol{\sigma} + \boldsymbol{\varepsilon}^t, \qquad \mathbf{M} = \mathbf{L}^{-1}, \qquad (A.3)$$

involve now vectors and matrices in place of second- and fourth-rank tensors, respectively.

The rotational transformation can conveniently be applied for the vector and matrix representations of second- and fourth-rank tensors using a 6×6 rotation matrix $\mathbf{Q}$. Its components are expressed in terms of the components Q_{ij} of the corresponding second-rank rotation tensor by, cf. Pedersen [99],

$$\mathbf{Q} = \begin{bmatrix} \mathbf{Q}_{NN} & \mathbf{Q}_{NS} \\ \mathbf{Q}_{SN} & \mathbf{Q}_{SS} \end{bmatrix}, \qquad (A.4)$$

where

$$\mathbf{Q}_{NN} = \begin{bmatrix} Q_{11}^2 & Q_{12}^2 & Q_{13}^2 \\ Q_{21}^2 & Q_{22}^2 & Q_{23}^2 \\ Q_{31}^2 & Q_{32}^2 & Q_{33}^2 \end{bmatrix},$$

$$\mathbf{Q}_{NS} = \sqrt{2} \begin{bmatrix} Q_{11}Q_{12} & Q_{11}Q_{13} & Q_{12}Q_{13} \\ Q_{21}Q_{22} & Q_{21}Q_{23} & Q_{22}Q_{23} \\ Q_{31}Q_{32} & Q_{31}Q_{33} & Q_{32}Q_{33} \end{bmatrix},$$

$$\mathbf{Q}_{SN} = \sqrt{2} \begin{bmatrix} Q_{11}Q_{21} & Q_{12}Q_{22} & Q_{13}Q_{23} \\ Q_{11}Q_{31} & Q_{12}Q_{32} & Q_{13}Q_{33} \\ Q_{21}Q_{31} & Q_{22}Q_{32} & Q_{23}Q_{33} \end{bmatrix},$$

$$\mathbf{Q}_{SS} = \begin{bmatrix} Q_{11}Q_{22} + Q_{21}Q_{12} & Q_{11}Q_{23} + Q_{21}Q_{13} & Q_{12}Q_{23} + Q_{22}Q_{13} \\ Q_{11}Q_{32} + Q_{31}Q_{12} & Q_{11}Q_{33} + Q_{31}Q_{13} & Q_{12}Q_{33} + Q_{32}Q_{13} \\ Q_{21}Q_{32} + Q_{31}Q_{22} & Q_{21}Q_{33} + Q_{31}Q_{23} & Q_{22}Q_{33} + Q_{32}Q_{23} \end{bmatrix}.$$

The rotation matrix $\mathbf{Q}$ is an orthogonal matrix, i.e. $\mathbf{Q}\mathbf{Q}^T = \mathbf{I}$. Now, for instance, the rotation of the strain vector $\boldsymbol{\varepsilon}$ and of the elastic stiffness matrix $\mathbf{L}$ is performed according to

$$\boldsymbol{\varepsilon}^* = \mathbf{Q}\boldsymbol{\varepsilon}, \qquad \mathbf{L}^* = \mathbf{Q}\mathbf{L}\mathbf{Q}^T, \qquad (A.5)$$

where vector $\boldsymbol{\varepsilon}^*$ and matrix $\mathbf{L}^*$ are the representations of the respective tensors given by $\varepsilon_{ij}^* = Q_{ik}Q_{jl}\varepsilon_{kl}$ and $L_{ijkl}^* = Q_{ip}Q_{jq}Q_{kr}Q_{ls}L_{pqrs}$.

A.2 Interior and Exterior Components

An intrinsic coordinate system associated with the interface with normal $\mathbf{n}$ is adopted in which the x_3-axis is perpendicular to the surface, i.e. parallel to the normal vector $\mathbf{n}$. In accord with the interior–exterior decomposition, cf. Sect. 2.2, the components of stress and strain vectors are rearranged, so that the interior (in-plane) $\boldsymbol{\sigma}_P$, $\boldsymbol{\varepsilon}_P$ and exterior (out-of-plane) $\boldsymbol{\sigma}_A$, $\boldsymbol{\varepsilon}_A$ components can be introduced:

$$\boldsymbol{\sigma} = \{\boldsymbol{\sigma}_P, \boldsymbol{\sigma}_A\} = \{\{\sigma_{11}, \sigma_{22}, \sqrt{2}\sigma_{12}\}, \{\sigma_{33}, \sqrt{2}\sigma_{13}, \sqrt{2}\sigma_{23}\}\},$$
$$\boldsymbol{\varepsilon} = \{\boldsymbol{\varepsilon}_P, \boldsymbol{\varepsilon}_A\} = \{\{\varepsilon_{11}, \varepsilon_{22}, \sqrt{2}\varepsilon_{12}\}, \{\varepsilon_{33}, \sqrt{2}\varepsilon_{13}, \sqrt{2}\varepsilon_{23}\}\}. \tag{A.6}$$

Note that $\boldsymbol{\sigma}_P$, $\boldsymbol{\sigma}_A$, $\boldsymbol{\varepsilon}_P$, and $\boldsymbol{\varepsilon}_A$ are now three-component subvectors. The components of the elastic moduli matrices $\mathbf{L}$ and $\mathbf{M}$ are rearranged accordingly, so that

$$\left\{ \begin{array}{c} \boldsymbol{\sigma}_P \\ \boldsymbol{\sigma}_A \end{array} \right\} = \left[\begin{array}{cc} \mathbf{L}_{PP} & \mathbf{L}_{PA} \\ \mathbf{L}_{AP} & \mathbf{L}_{AA} \end{array} \right] \left\{ \begin{array}{c} \boldsymbol{\varepsilon}_P - \boldsymbol{\varepsilon}_P^t \\ \boldsymbol{\varepsilon}_A - \boldsymbol{\varepsilon}_A^t \end{array} \right\},$$

$$\left\{ \begin{array}{c} \boldsymbol{\varepsilon}_P \\ \boldsymbol{\varepsilon}_A \end{array} \right\} = \left[\begin{array}{cc} \mathbf{M}_{PP} & \mathbf{M}_{PA} \\ \mathbf{M}_{AP} & \mathbf{M}_{AA} \end{array} \right] \left\{ \begin{array}{c} \boldsymbol{\sigma}_P \\ \boldsymbol{\sigma}_A \end{array} \right\} + \left\{ \begin{array}{c} \boldsymbol{\varepsilon}_P^t \\ \boldsymbol{\varepsilon}_A^t \end{array} \right\}, \tag{A.7}$$

where $\mathbf{L}_{PP}$, $\mathbf{M}_{PP}$, $\ldots$, are 3×3 submatrices of the elastic moduli matrices $\mathbf{L}$ and $\mathbf{M}$.

The intrinsic coordinate system is used in Sects. A.3 and A.4. Clearly, a rotational transformation, cf. (A.5), relates the components of vectors and matrices in the global and intrinsic coordinate systems. Therefore, though it is not explicitly visible, the operators derived below depend on the interface normal $\mathbf{n}$.

A.3 Interfacial Relationships

In the intrinsic coordinate system, the compatibility conditions take simple form, cf. (2.30),

$$\Delta\boldsymbol{\sigma}_A = \mathbf{0}, \qquad \Delta\boldsymbol{\varepsilon}_P = \mathbf{0}. \tag{A.8}$$

For given stress $\boldsymbol{\sigma}^+$ or strain $\boldsymbol{\varepsilon}^+$ on one side of the interface separating the two phases, the stress $\boldsymbol{\sigma}^-$ and strain $\boldsymbol{\varepsilon}^-$ on the other side can be found by solving the compatibility conditions (A.8) jointly with the constitutive relations (A.7) for each phase. The resulting interfacial relationships, cf. (2.33), are given by

$$\Delta\boldsymbol{\varepsilon} = -\mathbf{P}^0(\Delta\mathbf{L}\boldsymbol{\varepsilon}^+ - \Delta\boldsymbol{\sigma}^t), \qquad \Delta\boldsymbol{\sigma} = -\mathbf{S}^0(\Delta\mathbf{M}\boldsymbol{\sigma}^+ + \Delta\boldsymbol{\varepsilon}^t), \tag{A.9}$$

where $\boldsymbol{\sigma}^{t\pm} = \mathbf{L}^{\pm}\boldsymbol{\varepsilon}^{t\pm}$, and the operators $\mathbf{P}^0$ and $\mathbf{S}^0$ are given by

$$\mathbf{P}^0 = \left[\begin{array}{cc} \mathbf{0} & \mathbf{0} \\ \mathbf{0} & (\mathbf{L}_{AA}^-)^{-1} \end{array} \right], \qquad \mathbf{S}^0 = \left[\begin{array}{cc} (\mathbf{M}_{PP}^-)^{-1} & \mathbf{0} \\ \mathbf{0} & \mathbf{0} \end{array} \right]. \tag{A.10}$$

The following identities are trivially satisfied by $\mathbf{P}^0$ and $\mathbf{S}^0$,

$$\mathbf{P}^0 \mathbf{L}^- \mathbf{P}^0 = \mathbf{P}^0, \qquad \mathbf{S}^0 \mathbf{M}^- \mathbf{S}^0 = \mathbf{S}^0. \tag{A.11}$$

Note that the interfacial relationships (A.9) are valid for any smooth surface of discontinuity, not necessarily planar, provided that surface tension forces within the interface are negligible.

A.4 Macroscopic Elastic and Concentration Matrices

Consider now the simple laminate discussed in Sect. 2.6. Using the constitutive relations (A.7) and the compatibility conditions (A.8) together with the averaging rules

$$\Sigma = \{\sigma\} = \eta\sigma^- + (1 - \eta)\sigma^+, \quad \mathbf{E} = \{\varepsilon\} = \eta\varepsilon^- + (1 - \eta)\varepsilon^+, \tag{A.12}$$

where η is the volume fraction of the '$-$' phase, the effective macroscopic elastic matrices are obtained in the form

$$\begin{aligned}
\tilde{\mathbf{L}}_{uv} &= \eta\mathbf{L}_{uv}^- + (1 - \eta)\mathbf{L}_{uv}^+ \\
&\quad - \eta(1 - \eta)(\mathbf{L}_{uA}^+ - \mathbf{L}_{uA}^-)\mathbf{L}_{AA}^{*-1}(\mathbf{L}_{Av}^+ - \mathbf{L}_{Av}^-), \\
\tilde{\mathbf{M}}_{uv} &= \eta\mathbf{M}_{uv}^- + (1 - \eta)\mathbf{M}_{uv}^+ \\
&\quad - \eta(1 - \eta)(\mathbf{M}_{uP}^+ - \mathbf{M}_{uP}^-)\mathbf{M}_{PP}^{*-1}(\mathbf{M}_{Pv}^+ - \mathbf{M}_{Pv}^-),
\end{aligned} \tag{A.13}$$

where indices u, v denote either interior (P) or exterior (A) components, and

$$\begin{aligned}
\mathbf{L}_{AA}^* &= (1 - \eta)\mathbf{L}_{AA}^- + \eta\mathbf{L}_{AA}^+, \\
\mathbf{M}_{PP}^* &= (1 - \eta)\mathbf{M}_{PP}^- + \eta\mathbf{M}_{PP}^+.
\end{aligned} \tag{A.14}$$

The strain and stress concentration matrices, $\mathbf{A}^\pm$ and $\mathbf{B}^\pm$, respectively, are given by

$$\mathbf{A}^\pm = \begin{bmatrix} \mathbf{1} & \mathbf{0} \\ \mathbf{A}_{AP}^\pm & \mathbf{A}_{AA}^\pm \end{bmatrix}, \qquad \mathbf{B}^\pm = \begin{bmatrix} \mathbf{B}_{PP}^\pm & \mathbf{B}_{PA}^\pm \\ \mathbf{0} & \mathbf{1} \end{bmatrix}, \tag{A.15}$$

where

$$\begin{aligned}
\mathbf{A}_{AP}^- &= (1 - \eta)\,\mathbf{L}_{AA}^{*-1}\,(\mathbf{L}_{AP}^+ - \mathbf{L}_{AP}^-), \\
\mathbf{A}_{AP}^+ &= \eta\,\mathbf{L}_{PP}^{*-1}\,(\mathbf{L}_{AP}^- - \mathbf{L}_{AP}^+), \\
\mathbf{A}_{AA}^- &= \mathbf{L}_{AA}^{*-1}\,\mathbf{L}_{AA}^+, \\
\mathbf{A}_{AA}^+ &= \mathbf{L}_{AA}^{*-1}\,\mathbf{L}_{AA}^-,
\end{aligned}$$

and

$$\begin{aligned}
\mathbf{B}_{PP}^- &= \mathbf{M}_{PP}^{*-1}\,\mathbf{M}_{PP}^+, \\
\mathbf{B}_{PP}^+ &= \mathbf{M}_{PP}^{*-1}\,\mathbf{M}_{PP}^-, \\
\mathbf{B}_{PA}^- &= (1 - \eta)\,\mathbf{M}_{PP}^{*-1}\,(\mathbf{M}_{PA}^+ - \mathbf{M}_{PA}^-), \\
\mathbf{B}_{PA}^+ &= \eta\,\mathbf{M}_{PP}^{*-1}\,(\mathbf{M}_{PA}^- - \mathbf{M}_{PA}^+).
\end{aligned}$$

Finally, matrices $\mathbf{P}$ and $\mathbf{S}$ are given by

$$\mathbf{P} = \begin{bmatrix} \mathbf{0} & \mathbf{0} \\ \mathbf{0} & \mathbf{L}_{AA}^{*-1} \end{bmatrix}, \qquad \mathbf{S} = \begin{bmatrix} \mathbf{M}_{PP}^{*-1} & \mathbf{0} \\ \mathbf{0} & \mathbf{0} \end{bmatrix}. \qquad (A.16)$$

Note that, for $\eta = 0$, matrices $\mathbf{P}$ and $\mathbf{S}$ reduce to, respectively, $\mathbf{P}^0$ and $\mathbf{S}^0$ specified by (A.10).

B

Material Data of Selected Shape Memory Alloys

B.1 Transformation Strains and Microstructures

B.1.1 Cubic-to-Orthorhombic Transformation

The transformation strains of the six martensite variants in the cubic-to-orthorhombic transformation, such as the $\beta_1 \rightarrow \gamma_1'$ transformation in CuAlNi, are given by

$$
(\varepsilon_1^t)_{ij} = \begin{pmatrix} \alpha & 0 & \delta \\ 0 & \beta & 0 \\ \delta & 0 & \alpha \end{pmatrix}, \quad (\varepsilon_2^t)_{ij} = \begin{pmatrix} \alpha & 0 & -\delta \\ 0 & \beta & 0 \\ -\delta & 0 & \alpha \end{pmatrix},
$$

$$
(\varepsilon_3^t)_{ij} = \begin{pmatrix} \alpha & \delta & 0 \\ \delta & \alpha & 0 \\ 0 & 0 & \beta \end{pmatrix}, \quad (\varepsilon_4^t)_{ij} = \begin{pmatrix} \alpha & -\delta & 0 \\ -\delta & \alpha & 0 \\ 0 & 0 & \beta \end{pmatrix}, \quad \text{(B.1)}
$$

$$
(\varepsilon_5^t)_{ij} = \begin{pmatrix} \beta & 0 & 0 \\ 0 & \alpha & \delta \\ 0 & \delta & \alpha \end{pmatrix}, \quad (\varepsilon_6^t)_{ij} = \begin{pmatrix} \beta & 0 & 0 \\ 0 & \alpha & -\delta \\ 0 & -\delta & \alpha \end{pmatrix},
$$

relative to the cubic axes of the austenite. Parameters α, β, and δ are calculated from the lattice constants of the austenite and the martensite. The values of these parameters for the CuAlNi alloy are given in Table B.1 based on the lattice constants measured by Otsuka and Shimizu [93].

In the cubic-to-orthorhombic transformation in Cu-based alloys, the compatibility at the austenite–martensite interface is obtained by twinning. The crystallographic theory of martensite provides 96 distinct austenite–martensite microstructures, cf. Bhattacharya [13], Hane and Shield [40, 42]. For each variant pair (I, J), solution of the twinning equation (7.11) provides two twin plane normal vectors l. So-called compound twins are obtained for variant pairs $(1, 2)$, $(3, 4)$ and $(5, 6)$, for which, however, the habit plane equation (7.12) has no solution, i.e. a coherent stress-free austenite–martensite interface does not exist. Each of the remaining twelve variant pairs can form

Table B.1. Cubic-to-orthorhombic transformation in CuAlNi.

Transformation strain parameters, cf. (B.1)	
$\alpha = 0.0425$, $\beta = -0.0822$, $\delta = 0.0194$	

Austenite-twinned martensite microstructure, variant pair (1,3)		
Twin plane normal, l	$(0, 0.7071, -0.7071)$	Type I twins
	$(0.2155, 0.6905, 0.6905)$	Type II twins
Twin fraction, λ	0.3068	
Habit plane normal, $\mathbf{m}$	$(0.7330, 0.2009, -0.6499)$	
	$(0.6332, 0.2286, 0.7394)$	

one Type I twin and one Type II twin, for which austenite–martensite inter-faces exist. For each of these twins, solution of the habit plane equation (7.12) provides two twin fractions, $\lambda_1 = \lambda$ and $\lambda_2 = 1 - \lambda$, and for each of them two habit planes are found. There are thus $12 \times 2 \times 2 \times 2 = 96$ distinct solutions of (7.11) and (7.12). Importantly, in the geometrically linear theory, the four martensitic plates formed by the same variant pair and with the same twin fraction, though of different microstructure, i.e. having different twin and habit plane normals, have identical effective transformation strains, $\hat{\varepsilon}_\alpha^t = \lambda \varepsilon_I^t + (1 - \lambda) \varepsilon_J^t$.

The basic microstructural parameters of the austenite-twinned martensite interfaces in the CuAlNi alloy, following from the geometrically linear theory, are provided in Table B.1 for the (1,3) variant pair. These parameters fully define four microstructures, as four distinct combinations of l and $\mathbf{m}$ are possible. The remaining $96 - 4 = 92$ microstructures can be obtained by applying the rotations from the symmetry point group of cubic austenite. These rotations are listed, for instance, in Hane and Shield [40].

B.1.2 Cubic-to-Monoclinic Transformation

The monoclinic 6M martensites in Cu-based alloys have the "cubic axes" structure with a unique twofold axis along the edge of the original cubic unit cell, cf. Pitteri and Zanzotto [107]. The components of the transformation strain tensor of the first variant are given in the cubic basis by

$$(\varepsilon_1^t)_{ij} = \begin{pmatrix} \xi & 0 & 0 \\ 0 & \varrho & \sigma \\ 0 & \sigma & \tau \end{pmatrix}, \tag{B.2}$$

the transformation strains of the other variants are obtained by applying the rotations from the symmetry point group of the cubic austenite. Parameters ξ, ϱ, τ, and σ are calculated from the lattice constants of the austenite and martensite, cf. Stupkiewicz [131]. The values of these parameters for the 6M

(M18R) martensites in CuAlNi and CuZnAl alloys are given in Table B.2. The transformation strain parameters are computed using the lattice constants measured by Otsuka et al. [95] (for CuAlNi) and Chakravorty and Wayman [20] (for CuZnAl).

The monoclinic 6M martensites, such as the martensitic β_1' phase in CuAlNi, form internally faulted (untwinned) plates. The crystallographic theory of martensite provides 24 distinct austenite–martensite microstructures: for each martensite variant the habit plane equation (7.14) provides the shear magnitude k_{sf} and two habit plane normals $\mathbf{m}$. The habit planes and other microstructural parameters of the first variant are given in Table B.2. The habit plane normals and the shear systems of the remaining variants can be obtained by applying the rotations relating the variants. These rotations can be found in Pitteri and Zanzotto [107].

Table B.2. Cubic-to-monoclinic transformation in CuAlNi and CuZnAl.

Transformation strain parameters, cf. (B.2)		
	CuAlNi	CuZnAl
ξ	−0.0823	−0.0907
ϱ	0.1002	0.0838
τ	−0.0151	0.0105
σ	0.0194	0.0267

Austenite-martensite interface, variant 1		
	CuAlNi	CuZnAl
Shear magnitude, k_{sf}	0.03962	−0.00453
Habit plane normal, $\mathbf{m}$	$(0.701, -0.693, -0.167)$	$(0.700, -0.682, -0.212)$
	$(0.701, 0.693, 0.167)$	$(0.700, 0.682, 0.212)$
Shear direction, $\mathbf{s}$	$(0, 0.7071, 0.7071)$	
Shear plane normal, $\mathbf{n}$	$(0, -0.7071, 0.7071)$	

B.2 Elastic Constants

In general, for a specified alloy and a specified phase transformation, it is rather difficult to find in the literature the complete set of elastic constants, i.e. those of single crystals in the austenitic state and in the martensitic state. The available elastic constants of the CuAlNi and CuZnAl alloys are given in Tables B.3 and B.4, respectively.

Elastic constants of the cubic austenite of DO_3 type ordered structure in the CuAlNi alloy (β_1 phase) have been measured by Suezawa and Sumino [143]. Consistent data has recently been reported by Landa et al. [76]

Table B.3. Elastic constants of CuAlNi single crystals.

Austenite (β_1 phase)[a]			
c_{11}	c_{12}	c_{44}	(GPa)
142	126	96	

[a] Suezawa and Sumino [143]

2H martensite (γ_1' phase)[b]									
c_{11}	c_{22}	c_{33}	c_{44}	c_{55}	c_{66}	c_{12}	c_{13}	c_{23}	(GPa)
189	141	205	54.9	19.7	62.6	124	45.5	115	

[b] Yasunaga et al. [167]

Table B.4. Elastic constants of CuZnAl single crystals.

Austenite[a]			
c_{11}	c_{12}	c_{44}	(GPa)
130	118.4	86	

[a] Guenin et al. [36]

6M (M18R) martensite[b]												
c_{11}	c_{22}	c_{33}	c_{44}	c_{55}	c_{66}	c_{12}	c_{13}	c_{15}	c_{23}	c_{25}	c_{35}	c_{46} (GPa)
175	156	235	54	28	48	118	40	10	150	0	0	-10

[b] Rodriguez et al. [116]

who also measured the third order elastic constants (TOEC). While several martensitic phases are observed in the CuAlNi alloy, elastic properties of the orthorhombic γ_1' phase (2H structure) are only available, cf. Yasunaga et al. [167]. Similarly, in the case of the CuZnAl alloy, elastic constants of the cubic austenite (DO_3 structure) and of the monoclinic martensite (6M structure) are only available, cf. Guenin et al. [36] and Rodriguez et al. [116]. Note that in the case of both alloys, the measured elastic properties of the austenite and of the martensite refer to alloys of slightly different composition.

Due to the lack of the required elastic properties, only two martensitic transformations are characterized completely: the cubic-to-orthorhombic transformation in CuAlNi and the cubic-to-monoclinic transformation in CuZnAl. However, in the context of stress-induced transformations, the cubic-to-monoclinic ($\beta_1 \rightarrow \beta_1'$) transformation in CuAlNi is of great interest, as the β_1' phase is the typical stress-induced martensite in the popular CuAlNi alloy. In order to allow simulations of this transformation, the elastic constants of the monoclinic β_1' phase can be estimated using the elastic constants of the similar martensitic phase in CuZnAl. We note that both martensites have the same 6M (M18R) type structure, so one may assume that the elastic moduli tensors are, in a sense, similar. Such similarity is noticeable in the case of the austenitic phases of both alloys, both having the DO_3 structure, cf. Tables B.3

and B.4. It is thus assumed that the elastic constants of the monoclinic martensite in CuAlNi are equal to the corresponding constants in CuZnAl scaled by a constant C, namely

$$c_{ij}^{\text{CuAlNi}} = C\, c_{ij}^{\text{CuZnAl}}. \tag{B.3}$$

The scaling constant $C = 1.091$ is assumed as the mean ratio of the corresponding elastic constants of austenitic phases of CuAlNi and CuZnAl. Elastic constants of the monoclinic β_1' martensite in CuAlNi, estimated according to (B.3), are used in the numerical simulations in Sect. 8.3.2.

References

1. R. Abeyaratne and J.K. Knowles. On the driving traction acting on a surface of strain discontinuity in a continuum. *J. Mech. Phys. Solids*, 38:345–360, 1990.
2. R. Abeyaratne and J.K. Knowles. On the kinetics of an austenite→martensite phase transformation induced by impact in a Cu–Al–Ni shape-memory alloy. *Acta Mater.*, 45(4):1671–1683, 1997.
3. J. Aboudi. *Mechanics of Composite Materials*. Elsevier, Amsterdam, 1991.
4. M. Andrade, M. Chandrasekaran, and L. Delaey. The basal plane stacking faults in 18R martensite of copper base alloys. *Acta Metall.*, 32(10):1809–1816, 1984.
5. B. Avitzur and Y. Nakamura. Analytical determination of friction resistance as a function of normal load and geometry of surface irregularities. *Wear*, 107:367–383, 1986.
6. A. Azarkhin and O. Richmond. A model of ploughing by a pyramidal indenter – upper bound method for stress-free surfaces. *Wear*, 157:409–418, 1992.
7. J.M. Ball, C. Chu, and R.D. James. Hysteresis during stress-induced variant rearrangement. *J. Physique IV*, 5(C8):245–251, 1995.
8. J.M. Ball and R.D. James. Fine phase mixtures as minimizers of energy. *Arch. Ration. Mech. Anal.*, 100:13–50, 1987.
9. A.A. Bandeira, P. Wriggers, and P. de Mattos Pimenta. Numerical derivation of contact mechanics interface laws using a finite element approach for large 3D deformation. *Int. J. Num. Meth. Engng.*, 59:173–195, 2003.
10. N. Bay. Friction stress and normal stress in bulk metal forming processes. *J. Mech. Working Technol.*, 14:203–224, 1987.
11. K. Bhattacharya. Wedge-like microstructure in martensites. *Acta Metall. Mater.*, 39:2431–2444, 1991.
12. K. Bhattacharya. Comparison of the geometrically nonlinear and linear theories of martensitic transformation. *Continuum Mech. Thermodyn.*, 5:205–242, 1993.
13. K. Bhattacharya. *Microstructure of martensite: why it forms and how it gives rise to the shape-memory effect*. Oxford University Press, Oxford, 2003.
14. K. Bhattacharya and G. Dolzmann. Relaxed constitutive relations for phase transforming materials. *J. Mech. Phys. Solids*, 48:1493–1517, 2000.
15. K. Bhattacharya and R.D. James. A theory of thin films of martensitic materials with application to microactuators. *J. Mech. Phys. Solids*, 47:531–576, 1999.

16. F.P. Bowden and D. Tabor. *The Friction and Lubrication of Solids*. Clarendon Press, Oxford, 1953.

17. J.S. Bowles and J.K. MacKenzie. The crystallography of martensitic transformations I and II. *Acta Metall.*, 2:129–137 and 138–147, 1954.

18. J.L. Bucaille, E. Felder, and G. Hochstetter. Mechanical analysis of the scratch test on elastic and perfectly plastic materials with the three-dimensional finite element modeling. *Wear*, 249:422–432, 2001.

19. R. Buczkowski and M. Kleiber. A stochastic model of rough surfaces for finite element contact analysis. *Comp. Meth. Appl. Mech. Engng.*, 169:43–59, 1999.

20. S. Chakravorty and C.M. Wayman. Electron microscopy of internally faulted Cu–Zn–Al martensite. *Acta Metall.*, 25:989–1000, 1977.

21. J.M. Challen and P.L.B. Oxley. An explanation of the different regimes of friction and wear using asperity deformation models. *Wear*, 53:229–243, 1979.

22. C. Chu. *Hysteresis and microstructures: a study of biaxial loading on compound twins of copper-aluminum-nickel single crystals*. PhD thesis, University of Minnesota, 1993.

23. M.G. Cooper, B.B. Mikic, and M.M. Yovanovich. Thermal contact conductance. *Int. J. Heat Mass Transfer*, 12:279–300, 1969.

24. S.C. Cowin and M.M. Mehrabadi. Anisotropic symmetries of linear elasticity. *Appl. Mech. Rev.*, 48(5):247–285, 1995.

25. E.A. de Souza Neto, K. Hashimoto, D. Perić, and D.R.J. Owen. A phenomenological model for frictional contact accounting for wear effects. *Phil. Trans. R. Soc. Lond. A*, 354:819–843, 1996.

26. J. Dutkiewicz, H. Kato, S. Miura, U. Messerschmidt, and M. Bartsch. Structure changes during pseudoelastic deformation of CuAlMn single crystals. *Acta Mater.*, 44(11):4597–4609, 1996.

27. A. El Omri, A. Fennan, F. Sidoroff, and A. Hihi. Elastic-plastic homogenization for layered composites. *Eur. J. Mech. A/Solids*, 19:585–601, 2000.

28. J.D. Eshelby. Energy relations and the energy momentum tensor in continuum mechanics. In M.F. Kanninen et al., editors, *Inelastic Behaviour of Solids*, pages 77–114. McGrow-Hill, New York, 1970.

29. B. Feeny, A. Guran, N. Hinrichs, and K. Popp. A historical review on dry friction and stick-slip phenomena. *Appl. Mech. Rev.*, 51:321–341, 1998.

30. F. Feyel and J.-L. Chaboche. FE^2 multiscale approach for modelling the elasto-viscoplastic behaviouir of long fibre SiC/Ti composite materials. *Comp. Meth. Appl. Mech. Engng.*, 183:309–330, 2000.

31. A. Gałka, J.J. Telega, and R. Wojnar. Thermodiffusion in heterogeneous elastic solids and homogenization. *Arch. Mech.*, 46(3):267–314, 1994.

32. B.P. Gearing, H.S. Moon, and L. Anand. A plasticity model for interface friction: application to sheet metal forming. *Int. J. Plast.*, 17(2):237–271, 2001.

33. A.E. Giannakopoulos. The influence of initial elastic surface stresses on instrumented sharp indentation. *Trans. ASME J. Appl. Mech.*, 70:638–643, 2003.

34. J.A. Greenwood and J.B.P. Williamson. Contact of nominally flat surfaces. *Proc. R. Soc. Lond. A*, 295:300–319, 1966.

35. J.A. Greenwood and J.J. Wu. Surface roughness and contact: an apology. *Meccanica*, 36:617–630, 2001.

36. G. Guenin, M. Morin, P.F. Gobin, W. Dejonghe, and L. Delaey. Elastic constant measurements in β Cu–Zn–Al near the martensitic transformation temperature. *Scripta Metall.*, 11:1071–1075, 1977.

37. M.E. Gurtin. Two-phase deformations of elastic solids. *Arch. Ration. Mech. Anal.*, 84:1–29, 1983.

38. Z. Handzel-Powierża, T. Klimczak, and A. Polijaniuk. On the experimental verification of the Greenwood-Williamson model for the contact of rough surfaces. *Wear*, 154:115–124, 1992.

39. K.F. Hane. Bulk and thin film microstructures in untwinned martensites. *J. Mech. Phys. Solids*, 47:1917–1939, 1999.

40. K.F. Hane and T.W. Shield. Symmetry and microstructure in martensites. *Phil. Mag. A*, 78(6):1215–1252, 1998.

41. K.F. Hane and T.W. Shield. Microstructure in the cubic to monoclinic transition in titanium-nickel shape memory alloy. *Acta Mater.*, 47(9):2603–2617, 1999.

42. K.F. Hane and T.W. Shield. Microstructure in a cubic to orthorhombic transition. *J. Elasticity*, 59:267–318, 2000.

43. K.F. Hane and T.W. Shield. Microstructure in the cubic to trigonal transition. *Mater. Sci. Eng. A*, 291:147–159, 2000.

44. R. Hill. *The Mathematical Theory of Plasticity*. Oxford University Press, Oxford, U.K., 1950.

45. R. Hill. Elastic properties of reinforced solids: some theoretical principles. *J. Mech. Phys. Solids*, 11:357–372, 1963.

46. R. Hill. On constitutive macro-variables for heterogeneous solids at finite strain. *Proc. R. Soc. Lond. A*, 326:131–147, 1972.

47. R. Hill. Interfacial operators in the mechanics of composite media. *J. Mech. Phys. Solids*, 31(4):347–357, 1983.

48. H. Horikawa, S. Ichinose, K. Morii, S. Miyazaki, and K. Otsuka. Orientation dependence of $\beta_1 \rightarrow \beta_1'$ stress-induced martensitic transformation in a Cu–Al–Ni alloy. *Metall. Trans. A*, 19A:915–923, 1988.

49. Y. Huo and I. Müller. Nonequilibrium thermodynamics of pseudoelasticity. *Continuum Mech. Thermodyn.*, 5(3):163–204, 1993.

50. M.J. Hÿtch, Ph. Vermaut, J. Malarria, and R. Portier. Study of atomic displacement fields in shape memory alloys by high-resolution electron microscopy. *Mater. Sci. Eng. A*, 273–275:266–270, 1999.

51. H. Ike. Plastic deformation of surface asperities associated with bulk deformation of metal workpiece in contact with rigid tool. In M. Raous, M. Jean, and J.J. Moreau, editors, *Contact Mechanics*, pages 275–286. Plenum Press, New York, 1995.

52. R.D. James. Finite deformation by mechanical twinning. *Arch. Ration. Mech. Anal.*, 77:143–176, 1981.

53. R.D. James, R.V. Kohn, and T.W. Shield. Modeling of branched needle microstructures at the edge of a martensite laminate. *J. Physique IV*, C8:253–259, 1995.

54. Q. Jiang and H. Xu. Microobservation of stress induced martensitic transformation in CuAlNi single crystals. *Acta Metall. Mater.*, 40(4):607–613, 1992.

55. K.L. Johnson. The correlation of indentation experiments. *J. Mech. Phys. Solids*, 18:115–126, 1970.

56. K.L. Johnson. *Contact Mechanics*. Cambridge University Press, 1985.

57. K.L. Johnson. The application of shakedown principles in rolling and sliding contact. *Eur. J. Mech. A/Solids*, 11:155–172, 1992.

58. H. Kato, J. Dutkiewicz, and S. Miura. Superelasticity and shape memory effect in Cu–23at.%Al–7at.%Mn alloy single crystals. *Acta Metall. Mater.*, 42(4):1359–1365, 1994.

59. A.G. Khachaturyan. Some questions concerning the theory of phase transformations in solids. *Fiz. Tverd. Tela*, 8(9):2709–2717, 1966. English translation: *Sov. Phys. Solid State*, 8:2163–2168, 1967.

60. A.G. Khachaturyan. *Theory of Structural Transformations in Solids*. John Wiley and Sons, New York, 1983.

61. Y. Kimura and T.H.C. Childs. Surface asperity deformation under bulk plastic straining conditions. *Int. J. Mech. Sci.*, 41:283–307, 1999.

62. A. Klarbring. Derivation and analysis of rate boundary-value problems of frictional contact. *Eur. J. Mech. A/Solids*, 9(1):53–85, 1990.

63. R.V. Kohn. The relaxation of a double well energy. *Continuum Mech. Thermodyn.*, 3:193–236, 1991.

64. R.V. Kohn and S. Müller. Branching of twins near an austenite–twinned-martensite interface. *Phil. Mag. A*, 66(5):697–715, 1992.

65. K. Komvopoulos, N. Saka, and N.P. Suh. The mechanism of friction in boundary lubrication. *Trans. ASME J. Tribol.*, 107:452–462, 1985.

66. J. Korelc. Automatic generation of numerical codes with introduction to AceGen 4.0 symbolic code generator. Available at http://www.fgg.uni-lj.si/Symech/, 2000.

67. J. Korelc. Computational Templates. User manual. Available at http://www.fgg.uni-lj.si/Symech/, 2000.

68. J. Korelc. Multi-language and multi-environment generation of nonlinear finite element codes. *Engineering with Computers*, 18:312–327, 2002.

69. D.A. Korzekwa, P.R. Dawson, and W.R.D. Wilson. Surface asperity deformation during sheet forming. *Int. J. Mech. Sci.*, 34(7):521–539, 1992.

70. V. Kouznetsova, W.A.M. Brekelmans, and F.P.T. Baaijens. An approach to micro-macro modelling of heterogeneous materials. *Comp. Mech.*, 27:37–48, 2001.

71. I.V. Kragelsky, M.N. Dobychin, and V.S. Kombalov. *Friction and Wear – Calculation Methods*. Pergamon Press, Oxford, 1982.

72. P. Krasniuk and S. Stupkiewicz. Evolution of real contact area in the presence of bulk plastic deformation: the effect of strain hardening. Internal Report ENLUB/2002/1, IPPT PAN, 2002.

73. L. Krstulović-Opara, P. Wriggers, and J. Korelc. A C^1-continuous formulation for 3D finite deformation frictional contact. *Comp. Mech.*, 29(1):27–42, 2002.

74. S. Kucharski, T. Klimczak, A. Polijaniuk, and J. Kaczmarek. Finite element model for the contact of rough surfaces. *Wear*, 177:1–13, 1994.

75. D.C. Lagoudas, P.B. Entchev, P. Popov, E. Patoor, L.C. Brinson, and X. Gao. Shape memory alloys, Part II: Modeling of polycrystals. *Mech. Mater.*, 38:430–462, 2006.

76. M. Landa, V. Novák, P. Sedlák, and P. Šittner. Ultrasonic characterization of Cu–Al–Ni single crystals lattice stability in the vicinity of the phase transition. *Ultrasonics*, 42:519–526, 2004.

77. Y.-H. Lee and D. Kwon. Estimation of biaxial surface stress by instrumented indentation with sharp indenters. *Acta Mater.*, 52:1555–1563, 2004.

78. J. Lengiewicz, S. Stupkiewicz, J. Korelc, and T. Rodic. DDM-based sensitivity analysis and optimization for smooth contact formulations. In P. Wriggers and U. Nackenhorst, editors, *Analysis and Simulation of Contact Problems*, volume 27 of *Lecture Notes in Applied and Computational Mechanics*, pages 79–86. Springer, 2006.

79. C. Lexcellent, B.C. Goo, Q.P. Sun, and J. Bernardini. Characterization, thermomechanical behaviour and micromechanical-based constitutive model of shape-memory Cu–Zn–Al single crystals. *Acta Mater.*, 44(9):3773–3780, 1996.

80. D.Z. Liu and D. Dunne. Atomic force microscope study of the interface of twinned martensite in copper-aluminium-nickel. *Scripta Mater.*, 48(12):1611–1616, 2003.

81. R. Luciano and J.R. Willis. Boundary-layer corrections for stress and strain fields in randomly heterogeneous materials. *J. Mech. Phys. Solids*, 51:1075–1088, 2003.

82. G. Maciejewski, S. Stupkiewicz, and H. Petryk. Elastic micro-strain energy at the austenite–twinned martensite interface. *Arch. Mech.*, 57(4):277–297, 2005.

83. H. Morawiec. Stopy z pamięcią kształtu i ich zastosowanie. In W.K. Nowacki, editor, *Podstawy termomechaniki materiałów z pamięcią kształtu*, pages 7–54. IPPT PAN, Warszawa, 1996.

84. Z. Mróz and S. Stupkiewicz. Constitutive model of adhesive and ploughing friction in metal forming processes. *Int. J. Mech. Sci.*, 40:281–303, 1998.

85. I. Müller and H. Xu. On the pseudo-elastic hysteresis. *Acta Metall. Mater.*, 39(3):263–271, 1991.

86. V. Novák, P. Šittner, D. Vokoun, and N. Zárubová. On the anisotropy of martensitic transformations in Cu-based alloys. *Mater. Sci. Eng. A*, 273–275:280–285, 1999.

87. V. Novák, P. Šittner, and N. Zárubová. Anisotropy of transformation characteristics of Cu-base shape memory alloys. *Mater. Sci. Eng. A*, 234–236:414–417, 1997.

88. J.T. Oden and J.A.C. Martins. Models and computational methods for dynamic friction phenomena. *Comp. Meth. Appl. Mech. Engng.*, 52:527–634, 1985.

89. D.D. Olsson. *Limits of lubrication in sheet metal forming of stainless steel.* PhD thesis, Technical University of Denmark, October 2003.

90. L. Orgeas and D. Favier. Stress-induced martensitic transformation of a NiTi alloy in isothermal shear, tension and compression. *Acta Mater.*, 46(15):5579–5591, 1998.

91. K. Otsuka, T. Ohba, M. Tokonami, and C.M. Wayman. New description of long period stacking order structures of martensites in β-phase alloys. *Scripta Metall. Mater.*, 29:1359–1364, 1993.

92. K. Otsuka, H. Sakamoto, and K. Shimizu. Successive stress-induced martensitic transformations and associated transformation pseudoelasticity in Cu–Al–Ni alloys. *Acta Metall.*, 27:585–601, 1979.

93. K. Otsuka and K. Shimizu. Morphology and crystallography of thermoelastic Cu–Al–Ni martensite analyzed by the phenomenological theory. *Trans. Jap. Inst. Metals*, 15:103–108, 1974.

94. K. Otsuka and C.M. Wayman, editors. *Shape Memory Materials*. Cambridge University Press, 1998.

95. K. Otsuka, C.M. Wayman, K. Nakai, H. Sakamoto, and K. Shimizu. Superelasticity effects and stress-induced martensitic transformations in Cu–Al–Ni alloys. *Acta Metall.*, 24:207–226, 1976.

96. E. Patoor, A. Eberhardt, and M. Berveiller. Micromechanical modelling of superelasticity in shape memory alloys. *J. Physique IV*, C1:277–292, 1996.

97. E. Patoor, D.C. Lagoudas, P.B. Entchev, L.C. Brinson, and X. Gao. Shape memory alloys, Part I: General properties and modeling of single crystals. *Mech. Mater.*, 38:391–429, 2006.

98. V.J. Pauk and C. Woźniak. Plane contact problem for a half-space with boundary imperfections. *Int. J. Sol. Struct.*, 36:3569–3579, 1999.

99. P. Pedersen. *Elasticity—Anisotropy—Laminates*. Technical University of Denmark, 1997.

100. B.N.J. Persson. *Sliding Friction*. Springer Verlag, Berlin, 1998.

101. H. Petryk. Slip line field solutions for sliding contact. In *Friction, Lubrication and Wear—Fifty Years On*, volume II, pages 987–994, London, 1987. Proc. Instn. Mech. Engrs.

102. H. Petryk. Macroscopic rate-variables in solids undergoing phase transformation. *J. Mech. Phys. Solids*, 46:873–894, 1998.

103. H. Petryk and S. Stupkiewicz. Micromechanical modelling of stress-induced phase transition in shape memory alloys. *Arch. Metall. Mater.*, 49(4):765–777, 2004.

104. H. Petryk, S. Stupkiewicz, and G. Maciejewski. Modelling of austenite/martensite laminates with interfacial energy effect. In P. Tong and Q.P. Sun, editors, *Proc. IUTAM Symp. on Size Effects on Material and Structural Behaviour at Micron- and Nano-scales*, pages 151–162. Springer, 2006.

105. B. Peultier, T. Ben Zineb, and E. Patoor. Macroscopic constitutive law of shape memory alloy thermomechanical behaviour. Application to structure computation by FEM. *Mech. Mater.*, 38:510–524, 2006.

106. G. Pietrzak and A. Curnier. Large deformation frictional contact mechanics: continuum formulation and augmented Lagrangian treatment. *Comp. Meth. Appl. Mech. Engng.*, 177(3–4):351–381, 1999.

107. M. Pitteri and G. Zanzotto. Generic and non-generic cubic-to-monoclinic transitions and their twins. *Acta Mater.*, 46(1):225–237, 1998.

108. E. Pruchnicki. Hyperelastic homogenized law for reinforced elastomer at finite strain with edge effects. *Acta Mech.*, 129:139–162, 1998.

109. E. Rabinowicz. *Friction and Wear of Materials*. John Wiley & Sons, Inc., New York, 1965.

110. B. Raniecki. Termomechanika pseudosprężystosci materiałów z pamięcią kształtu. In W.K. Nowacki, editor, *Podstawy termomechaniki materiałów z pamięcią kształtu*, pages 55–140. IPPT PAN, Warszawa, 1996.

111. B. Raniecki and Ch. Lexcellent. R_L-models of pseudoelasticity and their specifications for some shape memory alloys. *Eur. J. Mech. A/Solids*, 13:21–50, 1994.

112. B. Raniecki and Ch. Lexcellent. Thermodynamics of isotropic pseudoelasticity in shape memory alloys. *Eur. J. Mech. A/Solids*, 17:185–205, 1998.

113. B. Raniecki and K. Tanaka. On the thermodynamic driving force for coherent phase transformations. *Int. J. Engng Sci.*, 32:1845–1858, 1994.

114. J.R. Rice. Continuum mechanics and thermodynamics of plasticity in relation to microscale deformation mechanisms. In A.S. Argon, editor, *Constitutive Equations in Plasticity*, pages 23–79. MIT Press, Cambridge, Mass., 1975.

115. C. Rodriguez and L.C. Brown. The thermal effect due to stress-induced martensite formation in β-CuAlNi single crystals. *Metall. Trans. A*, 11A:147–150, 1980.

116. P.L. Rodriguez, F.C. Lovey, G. Guenin, J.L. Pelegrina, M. Sade, and M. Morin. Elastic constants of the monoclinic 18R martensite of a Cu–Zn–Al alloy. *Acta Metall. Mater.*, 41(11):3307–3310, 1993.

117. A.L. Roytburd. Martensitic transformation as a typical phase transformation in solids. In D. Seitz and D. Turnbull, editors, *Solid State Physics*, volume 33, pages 317–380. Academic Press, New York, 1978.

118. A.L. Roytburd. Modified Clausius-Clapeyron equation for phase transformation hysteresis in solids (in Russian). *Fiz. Tverd. Tela*, 25(1):33–40, 1983. English translation: *Sov. Phys. Solid State*, 25:17–21, 1983.

119. A.L. Roytburd. Thermodynamics of polydomain heterostructures. I. Effect of macrostresses. *J. Appl. Phys.*, 83(1):228–238, 1998.

120. A.L. Roytburd. Thermodynamics of polydomain heterostructures. II. Effect of microstresses. *J. Appl. Phys.*, 83(1):239–245, 1998.

121. A.L. Roytburd and M.N. Pankova. The influence of external stresses on the orientation of the habit plane and substructure of stress-induced martensite plates in iron-based alloys. *Fiz. Met. Metalloved.*, 59(4):769–779, 1985. English translation: *Phys. Met. Metall.*, 59:131–140, 1985.

122. A.L. Roytburd and J. Slutsker. Thermodynamic hysteresis of phase transformation in solids. *Physica B*, 233:390–396, 1997.

123. A.L. Roytburd and J. Slutsker. Deformation of adaptive materials. Part I. Constrained deformation of polydomain crystals. *J. Mech. Phys. Solids*, 47:2299–2329, 1999.

124. A.L. Roytburd and J. Slutsker. Deformatin of adaptive materials. Part III: Deformation of crystals with polytwin product phases. *J. Mech. Phys. Solids*, 49:1795–1822, 2001.

125. E. Sanchez-Palencia. Boundary layers and edge effects in composites. In E. Sanchez-Palencia and A. Zaoui, editors, *Homogenization Techniques for Composite Media*, volume 272 of *Lecture Notes in Physics*, pages 121–192. Springer, Berlin, 1987.

126. T.W. Shield. Orientation dependence of the pseudoelastic behavior of single crystals of Cu–Al–Ni in tension. *J. Mech. Phys. Solids*, 43:869–895, 1995.

127. J.C. Simo and T.J.R. Hughes. *Computational Inelasticity*. Springer-Verlag, New York, 1998.

128. V. Smyshlyaev and J.R. Willis. On the relaxation of a three-well energy. *Proc. R. Soc. Lond. A*, 455:779–814, 1999.

129. N. Sridhar, J.M. Rickman, and D.J. Srolovitz. Twinning in thin films—I. Elastic analysis. *Acta Mater.*, 44(10):4085–4096, 1996.

130. S. Stupkiewicz. Extension of the node-to-segment contact element for surface-expansion-dependent contact laws. *Int. J. Num. Meth. Engng.*, 50:739–759, 2001.

131. S. Stupkiewicz. The effect of stacking fault energy on the formation of stress-induced internally faulted martensite plates. *Eur. J. Mech. A/Solids*, 23(1):107–126, 2004.

132. S. Stupkiewicz. Micromechanics of contact and interphase layers. IFTR Reports 2/2005, Institute of Fundamental Technological Research, Warsaw, 2005.

133. S. Stupkiewicz, J. Korelc, M. Dutko, and T. Rodič. Shape sensitivity analysis of large deformation frictional contact problems. *Comp. Meth. Appl. Mech. Engng.*, 191(33):3555–3581, 2002.

134. S. Stupkiewicz and A. Marciniszyn. Modelling of asperity deformation in the thin-film hydrodynamic lubrication regime. In N. Bay, editor, *Proc. 2nd Int. Conf. on Tribology in Manufacturing Processes ICTMP2004*, pages 695–702, Nyborg, Denmark, June 2004.

135. S. Stupkiewicz and Z. Mróz. A model of third body abrasive friction and wear in hot metal forming. *Wear*, 231:124–138, 1999.

136. S. Stupkiewicz and Z. Mróz. Modelling of bulk deformation effects on real contact area and friction in metal forming processes. In *Proc. EUROMECH 435 Friction and Wear in Metal Forming*, pages 63–70, Valenciennes, France, June 2002.

137. S. Stupkiewicz and Z. Mróz. Phenomenological model of friction accounting for subsurface plastic deformation in metal forming. In J.A.C. Martins and M.D.P. Monteiro Marques, editors, *Contact Mechanics*, Solid Mechanics and its Applications, pages 179–186, Dordrecht, 2002. Kluwer Academic Publishers.

138. S. Stupkiewicz and Z. Mróz. Phenomenological model of real contact area evolution with account for bulk plastic deformation in metal forming. *Int. J. Plast.*, 19(3):323–344, 2003.

139. S. Stupkiewicz and H. Petryk. Finite-strain micromechanical model of stress-induced martensitic transformations in shape memory alloys. *Mater. Sci. Eng. A.* doi:10.1016/j.msea.2006.01.112 (in print).

140. S. Stupkiewicz and H. Petryk. Modelling of laminated micro-structures in stress-induced martensitic transformation. *J. Mech. Phys. Solids*, 50:2303–2331, 2002.

141. S. Stupkiewicz and H. Petryk. Micromechanical modelling of stress-induced martensitic transformation and detwinning in shape memory alloys. *Journal de Physique IV*, 115:141–149, 2004.

142. S. Stupkiewicz and P. Sadowski. Micromechanical analysis of deformation and temperature inhomogeneities within rough contact layers. In P. Wriggers and U. Nackenhorst, editors, *Analysis and Simulation of Contact Problems*, volume 27 of *Lecture Notes in Applied and Computational Mechanics*, pages 325–332. Springer, 2006.

143. M. Suezawa and K. Sumino. Behaviour of elastic constants in Cu-Al-Ni alloy in the close vicinity of M_s-point. *Scripta Metall.*, 10:789–792, 1976.

144. N.P. Suh and H.-C. Sin. The genesis of friction. *Wear*, 69:91–114, 1981.

145. Q.P. Sun, editor. *Mechanics of Martensitic Phase Transformation in Solids.* Kluwer Academic Publishers, 2002.

146. P.M. Suquet. Elements of homogenization for inelastic solid mechanics. In E. Sanchez-Palencia and A. Zaoui, editors, *Homogenization Techniques for Composite Media*, volume 272 of *Lecture Notes in Physics*, pages 193–278. Springer, Berlin, 1987.

147. M.P.F. Sutcliffe. Surface asperity deformation in metal forming processes. *Int. J. Mech. Sci.*, 30(11):847–868, 1988.

148. W. Szczepiński. *Wstęp do analizy procesów obróbki plastycznej.* PWN, Warszawa, 1967.

149. P. Thamburaja and L. Anand. Polycrystalline shape-memory materials: effect of crystallographic texture. *J. Mech. Phys. Solids*, 49:709–737, 2001.

150. K. Varadi, Z. Neder, and K. Friedrich. Evaluation of the real contact areas, pressure distributions and contact temperatures during sliding contact between real metal surfaces. *Wear*, 200:55–62, 1996.

151. S. Vedantam and R. Abeyaratne. A Helmholz free-energy function for a Cu–Al–Ni shape memory alloy. *Int. J. Non-Lin. Mech.*, 40:177–193, 2005.

152. P. Šittner. Private communication. 2004.

153. P. Šittner and V. Novák. Anisotropy of Cu-based shape memory alloys in tension/compression thermomechanical loads. *Trans. ASME J. Eng. Mat. Tech.*, 121:48–55, 1999.

154. P. Šittner and V. Novák. Anisotropy of martensitic transformations in modeling of shape memory alloy polycrystals. *Int. J. Plasticity*, 16:1243–1268, 2000.

155. P. Šittner, V. Novák, and N. Zárubová. Martensitic transformations in [001] CuAlZnMn single crystals. *Acta Mater.*, 46(4):1265–1281, 1998.

156. T. Wanheim, N. Bay, and A.S. Petersen. A theoretically determined model for friction in metal working processes. *Wear*, 28:251–258, 1974.

157. M.S. Wechsler, D.S. Lieberman, and T.A. Read. On the theory of the formation of martensite. *Trans. AIME J. Metals*, 197:1503–1515, 1953.

158. D.J. Whitehouse and J.F. Archard. The properties of random surfaces in contact. *Proc. R. Soc. Lond. A*, 316:97–121, 1970.

159. J.R. Willis. Variational and related methods for the overall properties of composites. In *Advances in Applied Mechanics*, volume 21, pages 1–78. Academic Press, New York, 1981.

160. W.R.D. Wilson. Friction models for metal forming in the boundary lubrication regime. *Trans. ASME J. Eng. Mat. Technol.*, 113:60–68, 1991.

161. W.R.D. Wilson and W.M. Lee. Mechanics of surface roughening in metal forming processes. *Trans. ASME J. Manuf. Sci. Engrg.*, 123(2):279–283, 2001.

162. W.R.D. Wilson and S. Sheu. Real area of contact and boundary friction in metal forming. *Int. J. Mech. Sci.*, 30(7):475–489, 1988.

163. S. Wolfram. *The Mathematica Book, 4th ed.* Wolfram Media/Cambridge University Press, 1999.

164. Cz. Wozniak. Refined macrodynamics of periodic structures. *Arch. Mech.*, 45:295–304, 1993.

165. P. Wriggers. *Computational Contact Mechanics*. Wiley, Chichester, 2002.

166. P. Wriggers, T. Vu Van, and E. Stein. Finite element formulation of large deformation impact-contact problems with friction. *Comp. Struct.*, 37:319–331, 1990.

167. M. Yasunaga, Y. Funatsu, S. Kojima, K. Otsuka, and T. Suzuki. Measurement of elastic constants. *Scripta Metall.*, 17:1091–1094, 1983.

168. X.Y. Zhang, Q.P. Sun, and S.W. Yu. A non-invariant plane model for the interface in CuAlNi single crystal shape memory alloys. *J. Mech. Phys. Solids*, 48:2163–2182, 2000.

169. O.C. Zienkiewicz and R.L. Taylor. *The Finite Element Method*. Butterworth-Heinemann, Oxford, 5th edition, 2000.

170. F. Zimmermann and M. Humbert. Determination of the habit plane characteristics in the β–α' phase transformation induced by stress in Ti–5Al–2Sn–4Zr–4Mo–2Cr–1Fe. *Acta Mater.*, 50:1735–1740, 2002.

Index

Printing: Krips bv, Meppel
Binding: Stürtz, Würzburg